M. David Burghardt

Know Your Diesel

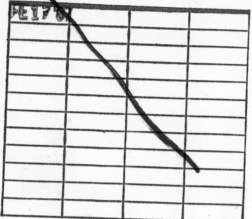
entice-Hall, Inc.
Englewood Cliffs, N. J. 07632

Library of Congress Cataloging in Publication Data

Burghardt, M. David.
   Know your diesel.

   Includes index.
    1. Diesel motor. I. Title.
TJ795.B797 1984   629.2′506   83-21269
ISBN 0-13-516591-1
ISBN 0-13-516584-9 (pbk.)

Editorial/production supervision and
  interior design: *Aliza Greenblatt*
Cover design: *Jeannette Jacobs*
Manufacturing buyer: *Anthony Caruso*

Printed in the United States of America

10 9 8 7 6 5 4 3 2 1

ISBN 0-13-516591-1 {C}

ISBN 0-13-516584-9 {P}

PRENTICE-HALL INTERNATIONAL, INC., *London*
PRENTICE-HALL OF AUSTRALIA PTY. LIMITED, *Sydney*
EDITORA PRENTICE-HALL DO BRASIL, LTDA., *Rio de Janeiro*
PRENTICE-HALL CANADA INC., *Toronto*
PRENTICE-HALL OF INDIA PRIVATE LIMITED, *New Delhi*
PRENTICE-HALL OF JAPAN, INC., *Tokyo*
PRENTICE-HALL OF SOUTHEAST ASIA PTE. LTD., *Singapore*
WHITEHALL BOOKS LIMITED, *Wellington, New Zealand*

To Linda

Contents

Preface

Diesel engines are increasingly being used as a power source in automobiles, trucks, and farming and construction equipment. They are more fuel efficient than their gasoline engine counterparts and have a greater life expectancy.

Everyone who purchases a diesel-powered vehicle needs to know how the diesel engine is made, what makes it run, and how well it performs. Whether you want to undertake engine repairs or just understand how the engine works, you need an information source that will readily and quickly educate you in knowing your diesel. This book is that source.

In addition, many people who already have a good understanding of gasoline engines are switching to diesel engines. If you do this, you will need new information to effectively operate, drive, or oversee the operation of diesel-driven vehicles because there are significant differences between gasoline and diesel engines. This book provides this information.

Using nearly 100 illustrations, the book clearly and concisely explains everything from counterweights to electronic fuel injection. This book goes significantly beyond your diesel maintenance manual, which does not explain why the parts are where they are, why they exist, and what their relationship is within the total engine. *Know Your Diesel* does. It is a perfect companion to your own engine's maintenance manual.

The book is arranged in five chapters. Chapter 1 covers the entire engine, its major parts, how they work, and how the engine performs. The next three chapters give greater detail and understanding to these parts. Chapter 5 contains a typical troubleshooting chart. At the end of the book is a glossary of commonly used words and phrases with their definitions.

It is my hope that this book will be very useful to you; it is written so that it would be. Many people and companies have been helpful to me in providing information and advice, such as Deutz, Caterpillar, and Mercedes-Benz. Cummins Engine Company has been particularly helpful.

M. David Burghardt

▐▌▌▐▌▐▌▐▌▐▌▐▌▐▌▐▌▐▌▌ ▐▌▐▌▐▌▐▌▐▌▐▌▐▌▐▌▌▐▌▐▌▐▌▐▌▐▌▐▌▐▌▐▌▐▌▐▌▐▌▐▌▐▌▐▌▐▌▐▌▐▌▌▐▌▐▌▐▌▐▌▐▌▐▌▐▌▐▌▐▌▐▌▐▌▐▌▌

How It's Built, How It Works

Diesel engines are increasingly being used as a power source in automobiles, trucks, and all sorts of heavy equipment vehicles. They are more efficient than their gasoline engine counterparts and have a longer operating life. What are the factors that make this engine, devised by Rudolf Diesel in 1892, different from the typical gasoline engine? The answers to these questions are developed throughout the book but are addressed specifically in this chapter. Rudolf Diesel originally tried to develop an engine that would run on powdered coal, but the engine exploded in the process. Switching to a liquid fuel enabled him to develop the engine which today we have in a highly evolved form. It is interesting to note that research is still under way on running a diesel engine on a mixture of coal and oil.

ENGINE PARTS

Let's get down to details. Figure 1.1 illustrates all the components that make up a diesel engine. Not all engines have this exact configuration, but all must have these components. It may seem a little overwhelming when looking at the illustration that all this will be understood. What we will do is analyze the engine systematically, understanding the role of each component. By the end of the book you will be able to refer back to this diagram and understand how all these parts relate and why they are included.

The biggest part in the engine is the *cylinder block*. Shown in Figure 1.1 is an in-line engine, where the cylinders form a single line. Another type of engine construction is that of a V, illustrated in Figure 1.2. In this case the cylinders form a V and are located in two banks, or rows. There are V-4, V-6, V-8, and V-12 engines, the number denoting the total number of cylinders. The cylinder block provides the framework, or structure, for holding all the other parts of the engine. It must be formed to withstand high temperature and pressures, which causes materials to loose their strength. The blocks are prepared from molds; molten metal is poured in and solidifies. The mold is removed, and the *casting,* as the block is now called, is heat treated to remove

1

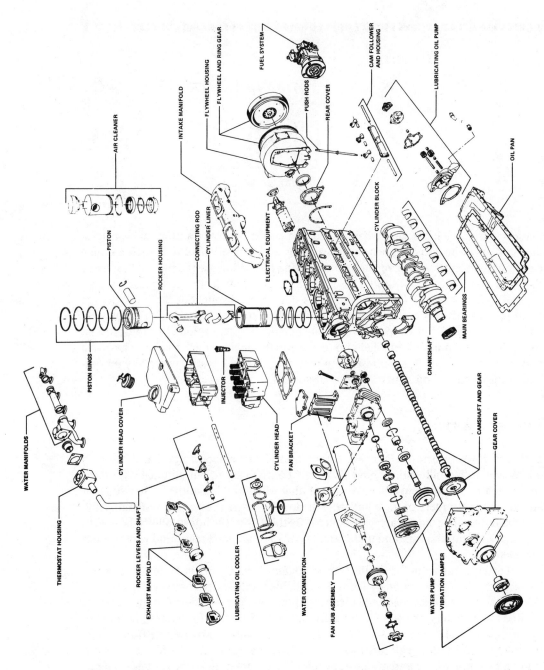

AIR CLEANER

INTAKE MANIFOLD

FLYWHEEL HOUSING

FLYWHEEL AND RING GEAR

PUSH RODS

REAR COVER

FUEL SYSTEM

CAM FOLLOWER AND HOUSING

LUBRICATING OIL PUMP

PISTON

CONNECTING ROD

CYLINDER LINER

ROCKER HOUSING

ELECTRICAL EQUIPMENT

CYLINDER BLOCK

OIL PAN

PISTON RINGS

INJECTOR

CRANKSHAFT

MAIN BEARINGS

WATER MANIFOLDS

CYLINDER HEAD COVER

CYLINDER HEAD

FAN BRACKET

CAMSHAFT AND GEAR

GEAR COVER

THERMOSTAT HOUSING

ROCKER LEVERS AND SHAFT

EXHAUST MANIFOLD

LUBRICATING OIL COOLER

WATER CONNECTION

FAN HUB ASSEMBLY

WATER PUMP

VIBRATION DAMPER

Figure 1.1 Disassembled diesel engine. (Reproduced with permission from Cummins Engine Co.)

2

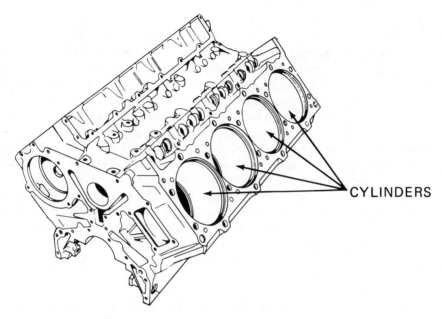

CYLINDERS

Figure 1.2 V-block construction. (Reproduced with permission from Cummins Engine Co.)

stresses that were formed during the metal's solidification process. In the heat-treating process, the casting is raised in temperature and allowed to cool very slowly. The casting is machined so that the openings meet the exact size specifications required by the parts fitting into the block.

The cylinders are the openings in the block that house the *pistons*. The cylinder head fits on top of the block and in conjunction with the piston forms the space in the cylinder where combustion occurs. Quite high temperatures and pressures (2000 to 3000°F and 700 psia) occur during the combustion process, which is why the cylinder block must be so strong. The pressures are greater than those encountered in a gasoline engine, so the parts in a diesel engine must be stronger to resist them. The engine must be cooled to remove some of the heat formed during the combustion process. This helps the metal keep its strength. Water passages are formed in the cylinder block to do just this. The piston is the only moving part in the cylinder and moves down under the influence of the high pressure in the cylinder. In doing so, it transmits to the *crankshaft* some of the energy released in the combustion process. This is the transforming of reciprocating motion into rotary motion. The rotary motion of the crankshaft is converted, through the transmission, into rotary motion at the wheels. This motion is also used to drive such auxiliary pieces of equipment as the generator, water pump, air conditioner, and fan.

Very often the cylinders have *liners,* which are metal sleeves that fit precisely into the cylinder. On high-speed engines, these liners are often called *sleeves.* The piston movement eventually will cause the surface it is rubbing against to wear. When the wear is too great, the engine will not perform correctly and other operational problems will occur. At this point, the liners can be replaced and the cylinder will be restored to its original dimensions. If liners were not used, there are two choices: bore the cylinder larger and replace the pistons with ones of larger diameter, or replace the entire cylinder block. The cylinder may be rebored only slightly, as it loses its side-wall strength and may crack under the high combustion temperatures and pressures. Liners are examined in more detail in Chapter 2. At this point we know why they are used—to absorb the wear that would otherwise affect the cylinder wall.

The *cylinder head* fits on top of the block. It houses the intake and exhaust valves and fuel injector. Of course, the head must contain passages for routing the air to the intake valves and directing the exhaust from the exhaust valves. The heads also have passages for cooling water and lubricating oil (Figure 1.3a). Because of the complex nature of the heads, they are formed from molds, just as was the cylinder block. Figure 1.3b shows a cylinder head for two cylinders; each cylinder may have its own head, typical of larger engines, or the entire bank may have a common head (Figure 1.3a shows the internal passages of the head). Additionally, the head must be machined to precise tolerances to house the valves and injectors, so that there will be no leakage of combustion gases from the cylinder.

The head contains the valves, but what tells the valves when to open and close? The *camshaft*. Cams, shown in Figure 1.4, are forged on a shaft. The shaft is directly driven by the crankshaft, by gears or by a chain. In either case there must be no slippage. Rocker arms are used to open the valves. In an overhead cam engine, shown in Figure 1.5, the nose of the cams push directly on the rocker arm to open the valves (the valves are held closed by springs). Sometimes the camshaft is located below the head in the block and is often gear-driven from the crankshaft. Figure 1.6 illustrates

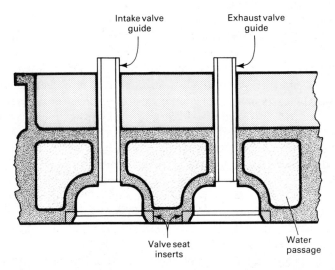

Intake valve
guide

Exhaust valve
guide

Valve seat
inserts

Water
passage

Figure 1.3   (a) Cross-sectional view of a cylinder head. (b) Cylinder head for a six-cylinder engine. (Reproduced with permission from Cummins Engine Co.)

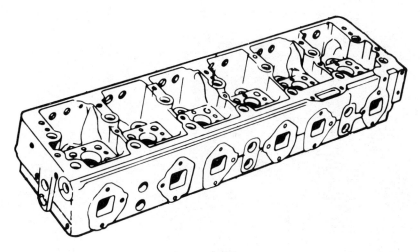

Figure 1.3(b)

this. The cam moves a pushrod, which moves the rocker arm, which in turn opens the valve.

   Fuel injectors are also located in the head. The injectors may be operated by the camshaft or by high fuel oil pressure. The unit injector system is most common, where a fuel pump receives the oil from the tank and delivers it continuously to the unit injectors. The fuel passes through the injector, cooling the injector when the injector is closed, and after being further pressurized by the injector when it is open, entering the combustion chamber. The oil that flows through the injector returns via a common line to the fuel tank. Figure 1.7 illustrates such a system schematically. Notice the use of the fuel filters to prevent dirt, metal, and foreign material from entering the injectors and scoring the finely machined parts. Each injector also has a small filter located within it.

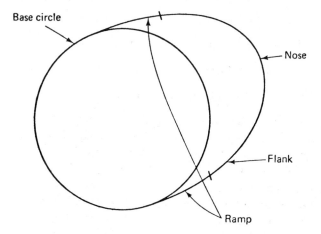

Figure 1.4   Schematic of a cam.

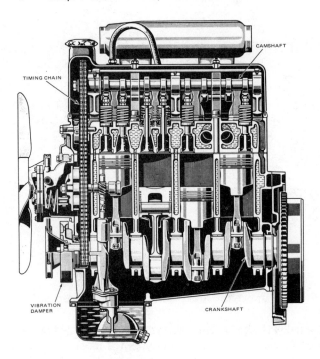

TIMING CHAIN

CAMSHAFT

VIBRATION
DAMPER

CRANKSHAFT

Figure 1.5 Overhead cam engine. (Reproduced with permission from Mercedes-Benz of North America.)

ENGINE CYCLES

What makes an engine run? We are now familiar with some of the major constituents, but simply knowing how they operate does not explain how the engine produces power. The source of the power is the fuel that is burned, and the transformation of the fuel's chemical energy into vehicle motion proceeds as follows:

Chemical energy	Fuel is supplied to engine.
Thermal energy	Combustion raises air–fuel temperature and pressure.
Mechanical energy	Piston moves because of the high pressure.
Shaft energy	Crankshaft rotates after translating piston motion.
Vehicle energy	Wheels convert crankshaft rotary motion.

The combustion process is one way in which a diesel engine is different from a gasoline engine. In the diesel's case, air is compressed by vertical piston movement, raising the air's temperature. The temperature must be great enough for fuel to ignite immediately upon entering the combustion chamber from the injector. No spark plugs are used as is the case in the gasoline engine.

Figure 1.8 shows a cylinder with its nomenclature. Notice that the piston will operate between two limits, the top-most part of its stroke, called *top dead center*

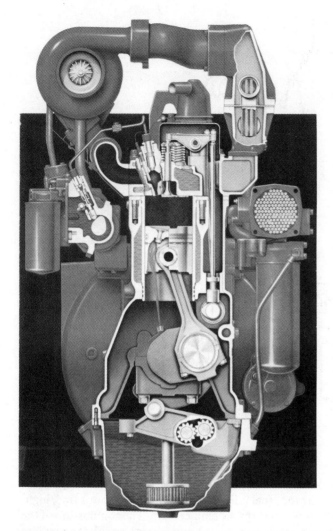

Figure 1.6 Engine with a rocker arm activated by a pushrod.

(TDC); and the bottom-most part of its stroke, called *bottom dead center* (BDC). The volume left between the piston crown (top) at TDC and the head is called the *cylinder clearance volume*. The *compression ratio, r,* is defined as the cylinder volume at TDC divided by the cylinder volume at BDC, or

$$r = \frac{\text{volume at TDC}}{\text{volume at BDC}}$$

The greater the compression ratio, the more the air is compressed and the greater will be the air's temperature and pressure. For diesel engines the compression ratio will vary from 14:1 to 24:1.

There are two *strokes* per engine revolution, one up and one down. Engine cycles are defined by the number of strokes that occur before an event repeats itself.

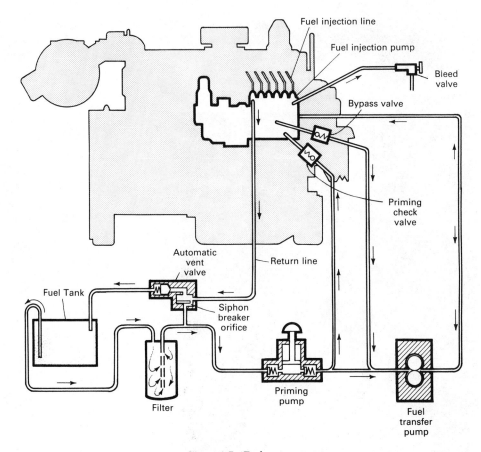

Figure 1.7 Fuel system.

Let us consider the four-stroke cycle illustrated in Figure 1.9. Near the top of the compression stroke, ideally at the top, fuel is injected into the cylinder in a fine, high-pressure spray and ignites because of the high air temperature. The combustion of the fuel continues as the piston moves downward on the power stroke, with the pressure ideally remaining constant, but actually increasing, then decreasing once combustion stops. The four strokes comprising the cycle may be classified as follows:

1.   *Intake stroke:* The intake valve is open; the exhaust valve is initially open, then it closes and the piston moves down, bringing a fresh-air mixture into the cylinder.

2.   *Compression stroke:* Both intake and exhaust valves are closed and the air is compressed by the upward piston movement.

3.   *Power stroke:* Both intake and exhaust valves are closed; injection and combustion occur, with the resultant pressure increase forcing the piston downward.

4.   *Exhaust stroke:* The exhaust valve is open, the intake valve is closed, and the

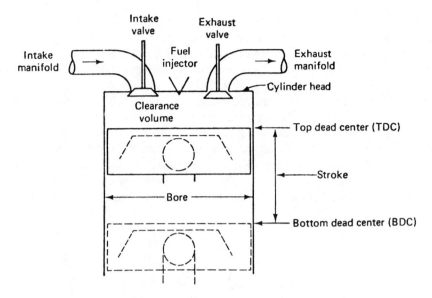

Figure 1.8 Schematic of major cylinder components and dimensions.

upward movement of the piston forces the products of combustion (exhaust) from the engine.

Since not all products of combustion are removed from the cylinder on the exhaust stroke because the piston goes only to TDC, it does not hit the head, there is a dilution of the incoming air charge by the remaining exhaust products. The greater the clearance volume, the greater the dilution.

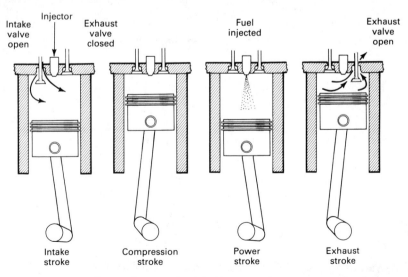

Figure 1.9 Four-stroke cycle: (a) intake; (b) compression; (c) power; (d) exhaust.

If there were just one diesel cycle, the number of revolutions per power stroke would be interesting but a cycle would not be defined by it. There is another diesel cycle, a *two-stroke cycle,* that does not use intake valves, and in some cases does not use exhaust valves. Air is supplied to the cylinder under a slight pressure. Located around the periphery of the cylinder are *ports,* which when uncovered by the piston on the downward stroke allow the air to enter the cylinder. Figure 1.10a shows a two-stroke-cycle diesel with the piston at bottom dead center. Air is entering the cylinder through the intake ports. A positive-displacement blower, driven from the crankshaft via gears, supplies the air. The exhaust valves open just before the piston top reaches the ports, so that the higher-pressure/higher-temperature products of combustion will not go into the intake manifold. As the piston moves upward (Figure 1.10b), the ports are covered, the exhaust valves close, and the air is compressed until the piston

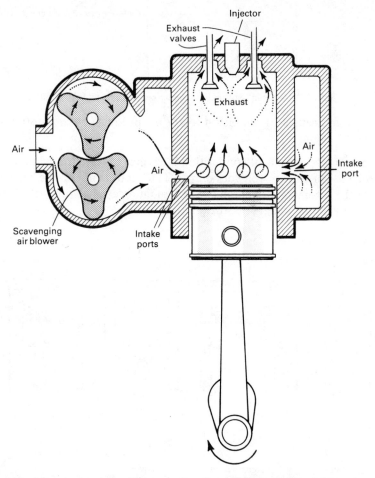

Figure 1.10 (a) Scavenging, exhaust, and intake on a two-stroke cycle. (b) Compression on a two-stroke cycle. (c) Combustion and power on a two-stroke cycle.

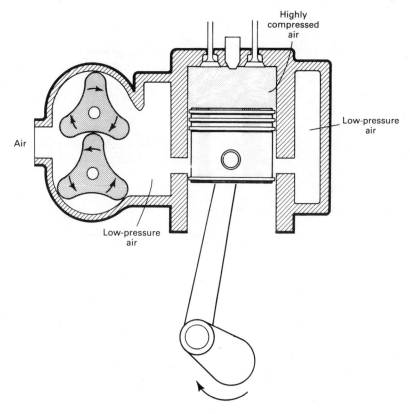

Figure 1.10(b)

reaches top dead center. Just before top dead center the injector admits fuel to the air and combustion begins (Figure 1.10c). The injection and combustion continue after top dead center, as the piston moves downward under the force of the high temperature and pressure of the products of combustion.

Regardless of the engine operation—on a two-stroke or a four-stroke cycle—the combustion process is similar. Figure 1.11 is a pressure-crank angle diagram. Before analyzing this diagram, let's look more closely at what is *crank angle*. Notice in Figure 1.10c that the piston is joined to the crankshaft with a connecting rod. In this picture the crankshaft is at an angle of almost 90 degrees from the vertical (TDC). This is the crank angle, 0 degrees at TDC, 180 degrees at BDC, returning to 360 degrees or 0 degrees at TDC. Now let's return to Figure 1.11. The air is compressed to 410 psi and about 900°F, and highly atomized fuel is injected (at point 1, before TDC) into this hot air. The fuel begins vaporizing and there is a period of time between the injection and the beginning of the combustion process, point 2. This is known as the *ignition delay*. Ignition delay occurs because the fuel must, first, vaporize, and second, have its temperature raised to the ignition point before combustion can occur. In a diesel engine, the combustion process starts throughout the

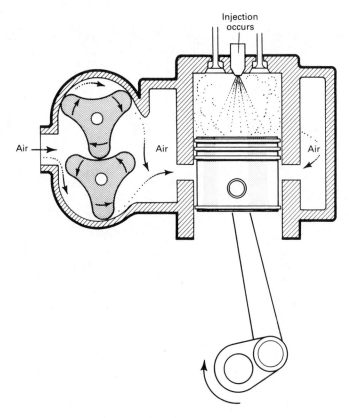

Figure 1.10(c)

entire combustion space, and the period between injection and ignition is called the *delay period*. Since fuel is still being injected during the delay period, it accumulates, and with the beginning of combustion at point 2, there is a rapid rise in temperature and pressure until point 3. By this time the accumulated fuel has been combusted and a controlled combustion process occurs until point 4. Here the fuel is burned immediately as it enters the cylinder. At point 4 injection stops and combustion should stop, although there is sometimes an afterburning period when unburned fuel continues burning. After combustion stops, the remaining pressure pushes the piston down. The dashed line shows how the pressure varies if no fuel is injected.

Figure 1.12a illustrates the four-stroke-cycle pressure–volume diagram. The area marked W_I represents the work produced by the engine. The W_{II} area represents the work consumed by the engine during the exhaust and intake strokes. The net work is the difference between the two areas. Area W_{II} has been exaggerated to show the effect of the exhaust and intake processes. There is some optimum value of exhaust valve opening which affects both areas. Point 4 represents the point where the exhaust valves open. Notice that if point 4 occurs earlier on the expansion stroke, less compressive work is required to push the exhaust gases out, as they flow out under

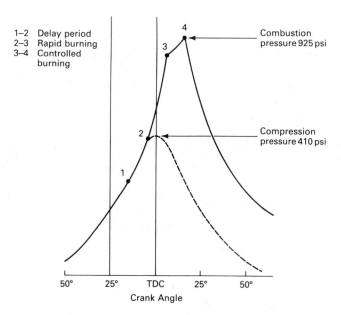

1–2 Delay period
2–3 Rapid burning
3–4 Controlled burning

Combustion pressure 925 psi

Compression pressure 410 psi

Crank Angle

Figure 1.11 Three phases to a combustion process.

their own pressure. This reduces the value of W_{II}. However, less work is done by the gases pushing the piston, so W_I is also reduced. Furthermore, if we think about the crank angle as the piston nears BDC, there is less force the connecting rod can exert on the rotating crankshaft. There is no rule as to how soon or late the exhaust valves must open; it varies with engine design. We will look at indicative values in just a moment.

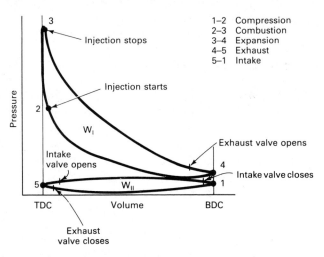

1–2 Compression
2–3 Combustion
3–4 Expansion
4–5 Exhaust
5–1 Intake

Injection stops

Injection starts

W_I

Exhaust valve opens

Intake valve opens

Intake valve closes

W_{II}

Exhaust valve closes

Pressure

TDC   Volume   BDC

Figure 1.12 Pressure–volume diagram of (a) a four-stroke cycle engine and (b) a nonsupercharged two-stroke cycle engine.

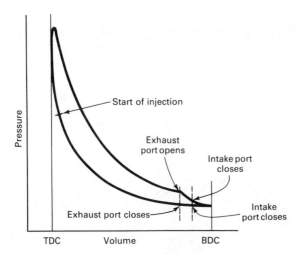

Figure 1.12(b)

The pressure–volume diagram for the two-stroke-cycle engine is illustrated in Figure 1.12b. The two-stroke-cycle engine has a power stroke every revolution. Thus intake, compression, power, and exhaust processes must occur within that revolution. To assist the exhaust process in the two-stroke-cycle engine, the engine is equipped with a scavenging blower or compressor, which raises the inlet air pressure to 2 to 5 psia above atmospheric pressure. Thus the intake air pushes the exhaust gas out. This is not the same as supercharging, which raises the inlet air pressure much higher. The work required to operate the scavenging air pump or compressor is charged against the engine, so the net work of the two-stroke-cycle engine is reduced by this amount. The scavenging pump must compress about 30 to 50 times the cylinder volume of air as the exhaust gas dilutes the incoming air charge and resists moving out of the cylinder. In addition to this loss of work, the combustion process is shortened by 20% and the compression stroke by about 12% compared to a four-stroke-cycle engine, to gain time for the scavenging process. Thus the net work of the two-stroke-cycle engine is not twice, but only one and one-half times that of a four-stroke-cycle engine.

A *valve-timing diagram,* or *polar timing diagram,* illustrates how the various processes in the engine cycles fit together with the valve openings and closings. Figure 1.13 illustrates a polar timing diagram for a two-stroke-cycle engine. The valve and port openings are indicative ones. Let's start at TDC on the power stroke. It proceeds to only 85 degrees after top dead center (ATDC), when the exhaust valves start to open. This allows the pressure to decrease before the intake ports are uncovered. In about 10 degrees the exhaust valves are fully open. The intake ports open at 135 ATDC and the air from the scavenging pump pushes the exhaust gases from the cylinder. This is called the *scavenging process.* At 45 degrees after bottom dead center (ABDC) the exhaust valves start to close, and at 50 degrees ABDC the intake ports close. The exhaust valve is still open, but closing, and the piston is moving upward. At 63 degrees ABDC the exhaust valves are fully closed and the compression process

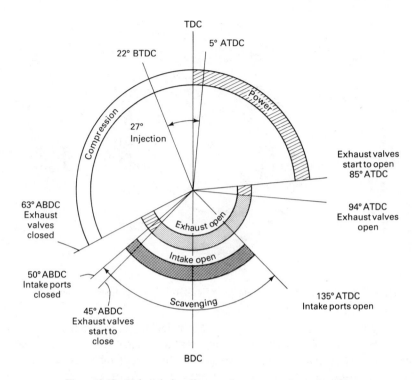

Figure 1.13 Valve–timing diagram for a two-stroke cycle engine.

starts. In some engines the exhaust valves close before the intake ports, which allows *supercharging* of the air in the cylinder. Supercharging means raising the pressure of the air in the cylinder above atmospheric pressure. Since there is more air, there is more oxygen and hence more fuel may be injected and burned, producing greater power. More on this later, as there are other performance factors to consider. Twenty-two degrees before top dead center (BTDC) the injection period begins and continues until 5 degrees ATDC. The pressure is high now and the piston starts again on the power stroke and the cycle repeats.

The four-stroke-cycle polar timing diagram is illustrated in Figure 1.14. In this case the cycle is 720 degrees, or two crankshaft rotations. Let's start at 28 degrees BTDC, where the intake valves open. The intake stroke proceeds until 35 degrees ABDC, when the valves close. Why does the intake valve stay open after bottom dead center? To provide a greater air mass to flow into the cylinder, even though the piston is moving upward. The air in the cylinder has less pressure than the air in the intake manifold even after the piston has passed bottom dead center. This is due to the inertia of the air mass entering the cylinder. Also, as the piston is moving down, it reduces the cylinder pressure below that of the intake manifold. Since there is very little vertical piston motion when the crank is crossing bottom dead center, compression is slight, and there is time for additional air to enter the cylinder. Compression of the air

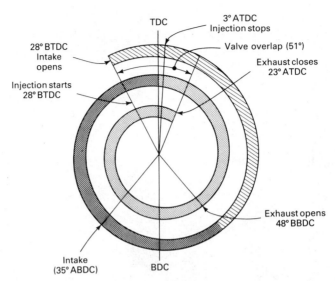

Figure 1.14 Valve–timing diagram for a four-stroke cycle engine with valve overlap.

now proceeds until top dead center. About 28 BTDC injection begins and continues until 3 degrees ATDC. The power stroke begins at top dead center and continues until the exhaust valves open at 48 degrees BBDC. The exhaust process continues until 23 ATDC. Notice that both the exhaust and intake valves are open from 28 degrees BTDC until 23 degrees ATDC. This period is called *valve overlap.* It allows the inlet air to sweep the exhaust gases from the clearance volume of the engine and permits minimum dilution of the air charge in the engine. The exhaust valve opens early to reduce the cylinder pressure and, hence, reduce the negative work required to push out the exhaust products. Any power that might be derived by delaying the exhaust valve opening is small, as there is little vertical motion in the piston-crank assembly and thus there is little torque benefit from the piston's acting on the crankshaft at this point.

The values given for valve openings were indicative of a high-speed diesel engine. The valves must open earlier than we might imagine to allow time for exhaust and intake processes. The absolute time for intake, exhaust, and injection nearly remains constant, so as the engine speed increases, the valves must open earlier to compensate for this.

COMBUSTION AND THE COMBUSTION CHAMBER

When thinking of an engine, very often we imagine the combustion process: fuel burning, exploding, and the piston moving. What factors cause the fuel to burn well? There are several, some of which are air temperature, fuel quality, air/fuel ratio, and turbulence of the burning fuel-air mixture. Others we'll examine later. We have already discovered that by raising the compression ratio, the temperature of the air during the compression process increases. The quality of the fuel is imposed by the type of fuel you buy. Chapter 3 examines fuel classification in detail.

What is the *air/fuel ratio,* and why is it important? Let's consider air first. We are interested in the oxygen in the air, as that is what combines with the fuel. Air has 21% oxygen by volume, the rest essentially being nitrogen, an inert gas that does not enter the chemical reaction. Thus for every liter of air brought into an engine, less than one-fourth is oxygen. When the fuel is injected into the cylinder, it mixes with the oxygen, reaches its ignition point, and burns. It is important that the fuel have enough oxygen to completely oxidize, or black soot particles will form. The air/fuel ratio is an indication of whether or not there is enough oxygen present. There is a theoretical minimum amount of air required to burn the fuel, about 15 parts of air to 1 part of fuel on a mass or weight basis, or an air/fuel ratio of 15:1. Practically, it is not possible to have an engine operate with this ratio, because to burn completely, the fuel must be provided with excess air.

Consider a small droplet of fuel as it is injected into the combustion chamber. It is surrounded by air and some of the fuel vaporizes and ignites. The products of combustion tend to block the remaining fuel's access to the remaining oxygen. If there is more than the minimum amount of oxygen, the remaining fuel can burn more readily.

Excess air is the percent above the theoretical minimum amount of air required for combustion. Thus an engine may run with 50% excess air or 50% more air than is theoretically required. The more air used, the leaner the combustion process is said to be, thus the greater the air/fuel ratio. The richer the mixture, the lower the air/fuel ratio, and the more fuel there is present. Sometimes at idle conditions, white smoke may appear. Because of the lower temperatures in the combustion chamber and reduced turbulence because of the reduced engine speed, the fuel does not burn. This is different from partially burned fuel, which causes black smoke. At full-load conditions, the engine is designed to use the maximum amount of fuel possible, with black smoke.

Engines are designed so that the air, and hence the products of combustion, swirl around the combustion chamber (that part of the cylinder where combustion occurs)—this is called *turbulence.* Assuring movement between the fuel, air, and products of combustion allows the fuel to come in contact with the available oxygen molecules. Combustion chambers are designed to promote turbulence, and several different types are on the market.

The problem with open combustion chambers is that they react more slowly to speed changes than other designs do. The change in speed causes an imbalance between the air and fuel requirements, the result of which is a smoky exhaust. The use of electronic means to control when to begin injection minimizes this. The types of combustion chambers used in high-speed engines are discussed next.

The most common and the most fuel-efficient combustion chamber is the *open combustion chamber.* Figure 1.15 shows the schematic of an open combustion chamber. The fuel injector supplies finely atomized fuel directly into the cylinder. Combustion occurs, with the resultant pressure buildup, forcing the piston down.

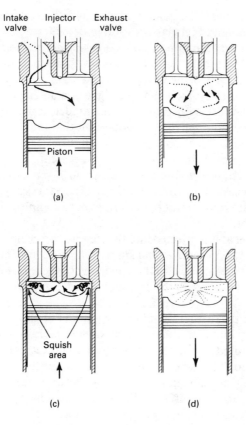

Figure 1.15 Increased air turbulence due to squish area action.

There is more to it, though. In Figure 1.15a, the intake air is given a swirling motion by grooves cut into the intake passage before the valve. Other engines have the intake ports at an angle to the valve, giving the air a rotation. Another method to achieve inlet air swirl is illustrated in Figure 1.16. The intake valve shown here is called a *masked valve*; it has a lip on its surface. The air is deflected by this mask, imparting a swirling motion. Note in Figure 1.15c, that the piston has two squish areas, where very little air remains at TDC. As the piston approaches TDC, the air is forced from the volume between the piston and head into the center cylinder volume—again add-

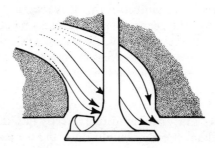

Figure 1.16 Increased air turbulence by a masked valve.

ing turbulence to the air, and in this case, the air–fuel mixture, as injection starts before top dead center. Figure 1.15d shows the injection process. There are operational difficulties encountered with open combustion chambers; often the injectors clog because of the very small openings in the multiorifice tip. A small amount of fuel is forced under high pressure through orifices, atomizing the fuel. These orifices can clog. The big advantage is greater fuel economy than that of the other combustion chamber types.

The *precombustion chamber* is illustrated in Figure 1.17. A portion of the clearance volume, 25 to 40%, is located in a separate chamber, the precombustion chamber. Air is forced through the small opening on the compression stroke, causing turbulence in the precombustion chamber. The injector supplies fuel to the air in the precombustion chamber; the pressure builds, forcing the burning fuel and air into the main chamber, where the combustion process concludes. The high velocity of the gases exiting the precombustion chamber assures good mixing and complete combustion. Compared to the open combustion chamber, the heat losses are greater; hence not as much thermal energy can be converted into work. The injector need not atomize the fuel as finely as in the open combustion chamber, and fuels of varying quality may be burned.

The *turbulence chamber* is illustrated in Figure 1.18. Virtually all of the clearance volume (80 to 90%) is located in the turbulence chamber. There air is given

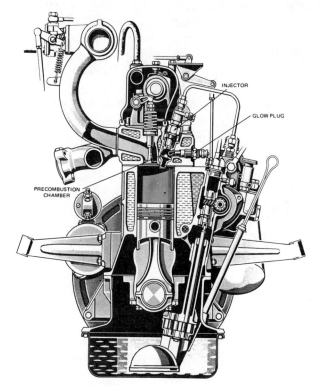

INJECTOR

GLOW PLUG

PRECOMBUSTION
CHAMBER

Figure 1.17 Engine using a precombustion chamber. (Reproduced with permission from Mercedes-Benz of North America.)

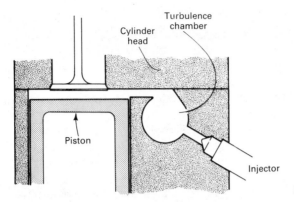

Figure 1.18 Turbulence chamber.

a swirling motion by the spherical shape of the chamber. Fuel is injected, combustion occurs, and high-pressure gases flow into the main chamber, where combustion is completed.

A *Lanova energy cell* is illustrated in Figure 1.19. It is a combination of a precombustion chamber and a turbulence chamber and is named for its developer. About 10% of the clearance volume is in the energy cell. The nozzle is located directly across from the energy cell and the combustion chamber is formed in a figure-eight shape. Figure 1.20 shows a top view of this type of chamber. Air moves in a swirling motion in both the main chamber and the energy cell on the compression stroke. Fuel is injected in a stream, so some of it will enter the energy cell. Most of it remains in the first portion of the cell, but enough enters the second, circular portion, for combustion to begin. The high pressure in this second chamber forces the fuel and air from the first triangular chamber into the main chamber. The figure-eight design assures turbulent mixing and the combustion proceeds in the main chamber in a smooth, steady fashion.

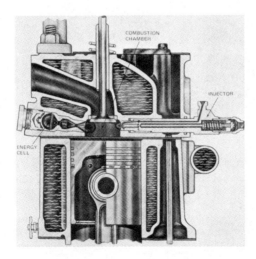

Figure 1.19 Energy cell located in a cylinder. (Reproduced with permission from Mercedes-Benz of North America.)

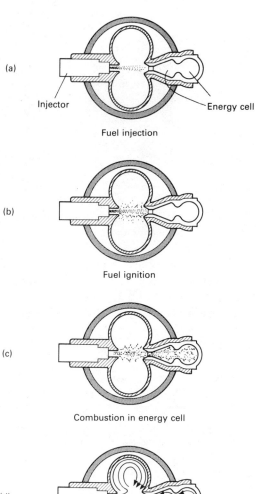

(a)

Injector

Energy cell

Fuel injection

(b)

Fuel ignition

(c)

Combustion in energy cell

(d)

Combustion in main chamber

Figure 1.20 Sequence of combustion in a Lanova energy cell combustion chamber viewed from the top: (a) fuel injection; (b) fuel ignition; (c) combustion in the energy cell; (d) combustion in the main chamber.

PERFORMANCE FACTORS

Now that we know how combustion occurs, and the sequence of the various processes, let's turn our attention to the combination of these events in making the engine perform. One of the most direct ways of understanding the engineering performance is to look at a *heat balance,* or energy inventory, of the engine. Figure 1.21 illustrates such a balance; this is indicative of the parameters involved. The energy value of the fuel is 100%; part of this is converted into work, part into friction, part

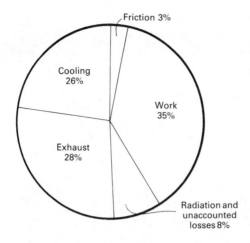

Figure 1.21 Energy division in a four-stroke-cycle engine.

into the cooling system, part into the exhaust, and the rest as radiation, plus unaccounted-for losses. The engine uses a certain amount of fuel and delivers an amount of power to the driveshaft. This is what we want, and we see in this case that 35% of the fuel's energy is converted into this energy form. Some of the fuel's energy is dissipated as friction in the bearings and gears. A large portion of the fuel's energy is removed by the cooling water. This is necessary or the metal parts could not withstand the high temperatures in the cylinder. Another large portion leaves as hot exhaust gases. Their being at a high temperature reflects the energy loss. The engine also loses heat by radiation to the surroundings. The last portion reflects inaccuracies in measurement techniques—hence the unaccounted-for losses. The reason the heat balance is useful is that the total energy pie remains constant—if the work decreases, something else must increase. In terms of troubleshooting, we look for this type of effect.

Let's examine next some of the factors that affect performance.

Engine Speed.   The faster the engine runs, the more frequently the power stroke occurs, and hence the more power is available.

Compression Ratio.   The higher the compression ratio, the greater will be the average pressure in the cylinder on the power stroke, which translates into more force pushing the piston down, and more power delivered to the crankshaft. Why not just increase the compression ratio? Such an increase in pressure requires stronger (heavier, for the most part) components (piston, block, bearings), and hence the compression ratio is limited by material considerations.

Air Mass.   On a naturally aspirated engine, one that pulls in the air by the piston moving down on the intake stroke, the flow increases with speed, but the mass of air in the cylinder may not be greater because the time for the air to get into the cylinder is less. The greater the air mass in the cylinder, the more fuel that may be burned, and the more power derived from the power stroke. For every self-aspirated engine there is one speed that yields the maximum air mass in the cylinder.

Injection Timing.   The fuel is injected to achieve the maximum pressure in the cylinder, without causing the engine to run roughly. Figure 1.22 illustrates the effect of the start of injection on the maximum pressure and the rate of pressure increase. The maximum pressure and pressure rise occur with injection 40 degrees BTDC. In this case there would be severe knocking, or small explosions of fuel in the combustion chamber—not a desired effect. At 60 degrees BTDC, the maximum pressures decreases, because the temperature was not sufficient to ignite the fuel properly. Also, the fuel will be not finely atomized due to droplets coalescing, and when the combustion does begin, and the process proceeds slowly. When late injection occurs, the piston is moving downward and the pressure rise is not as apparent. Note that when injection began at 10 degrees BTDC, the maximum pressure occurs significantly after top dead center.

Valve and Port Timing.   As discussed earlier when looking at pressure–volume diagrams, the exhaust and intake valve openings and closings are timed to give the greatest mass of fresh-air charge, while expelling the exhaust, with minimum work. Additionally, in the case of two-stroke-cycle engines, the exhaust valve (port) closing must be such as to minimize loss of scavenging air, called *blowdown.*

Let's see how the pressure–time and pressure–volume diagrams are affected by load. Figures 1.23 and 1.24 illustrate these and show the wisdom of operating the

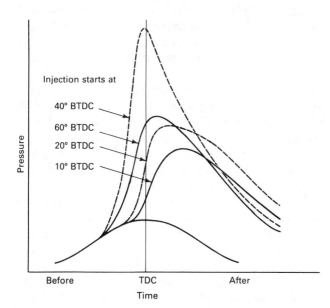

Figure 1.22 Illustration of how injection timing affects pressure rise during combustion.

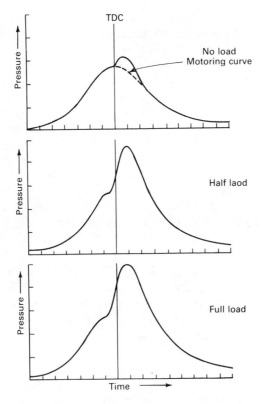

Figure 1.23 Changes in pressure–time diagrams at various engine loads.

engine at full-load conditions. Notice in Figure 1.23 the increase in peak pressure and the increase in the rate of pressure rise. This is manifested in Figure 1.24 by larger areas. The greater the area, the more work done per cycle. Of course, more fuel is burned, but the fixed losses (exhaust, cooling, friction) take a proportionally smaller bite of the fuel's energy.

Engines have performance curves which the manufacturer will provide. Figure 1.25 illustrates one such curve. Let's first find out what torque, brake specific fuel consumption (bsfc), and brake horsepower (bhp) are. Let's take the easiest first—*brake horsepower*. Each piston contributes a certain amount of work in moving down under high pressure (a force acting through a distance). This is energy and the work done in a given time is power (energy per unit time). For instance, the work is equal to 100 Btu (British thermal units, a measure of energy) per cycle, and there are 1000 power cycles per minute (a four-stroke engine running at 2000 rpm), so the power is 100×1000 or 100,000 Btu/min. This can be converted into units of power by conversion factors. It is equivalent to 128.7 horsepower (hp) or 175.9 kilowatts (kW). We can determine the engine's power in several locations—inside the cylinder, from analyzing the theoretical or ideal temperature and pressure variation (this is called *theoretical* or *ideal power*), or inside the cylinder, by determining the pressure–volume diagram (similar to Figure 1.24)—and calculating the work and

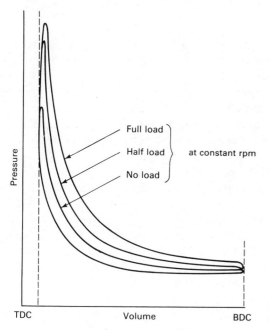

Figure 1.24 Changes in pressure–volume diagrams at various engine loads.

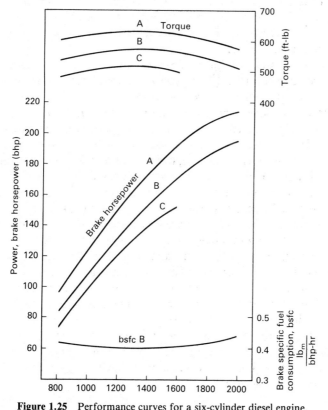

Figure 1.25 Performance curves for a six-cylinder diesel engine.

power the air–fuel mixture indicated it did in moving the piston (called indicated power), we can measure the power delivered by the crankshaft, called shaft or brake horsepower (shp, bhp).

Next, there is the *brake specific fuel consumption*. When the engine is operating, it burns a certain amount of fuel, let's say 6 gallons per hour. Gallons, unfortunately, are a volumetric measurement, 0.134 cubic feet. For No. 1D diesel fuel, there are 6.95 pounds of fuel in a gallon, so the engine in this case uses $6 \times 6.95 = 41.6$ pounds per hour. The brake power is measured and found to be 100 brake horsepower. If we divide 41.6 by 100, we get 0.416. What is this? It takes 0.416 lb of fuel per hour to produce each brake horsepower—this is the brake specific fuel consumption, written in symbols as 0.416 lb/bhp-hr. The lower this term is, the better, as it takes less fuel to produce a given horsepower.

Now what about *torque*? Consider that you are holding a wrench in your hand, trying to loosen a nut. When you pull on the wrench, you are exerting a tangential force, acting at a distance from the nut. This combination of tangential force acting at a distance is called torque. Torque is the capacity to do work, whereas *power* is the rate at which work may be done. Torque is measured in ft-lb$_f$. Let us consider a truck pulling a load. The torque of the engine will determine whether or not the truck is able to pull the load; the power will determine the rate, that is, how fast the load may be pulled. Back to the wrench. It requires a certain torque to turn the nut; how fast it can be turned depends on the power. Thus both torque and power are important considerations in selecting an engine.

Referring once again to Figure 1.25, we see that there are three curves, for the torque and power and the brake specific fuel consumption curve corresponding to the B curve. Curves A represent the maximum performance of the engine, found under laboratory conditions. The upper power limit is determined by smoky exhaust; the engine cannot convert more fuel into work. There is a maximum speed limitation on the engine as well. As curves B are developed for maximum conditions found under intermittent service, the engine is not expected to run at the maximum value for a sustained period of time. This would be indicative of automotive and truck engines, which would not be run continuously at high speeds. For machines that run continuously, curves C reflect the engine performance. In this case the maximum speed is limited to providing a safety factor to assure that the engine does not fail.

Notice the relationship between bsfc and power. As the engine speed increases the bsfc decreases, goes through a minimum, and then increases. This reflects less time in which the combustion process can occur, so not all the fuel is burned. There is a minimum, however, and a corresponding maximum on the torque curve. Why? In self-aspirated engines, there will be one speed where the mass of air drawn into the cylinder is a maximum. This will yield the highest average pressure in the cylinder because the largest amount of oxygen is present for the combustion process. The torque is proportional to this pressure; hence it is a maximum and the fuel is used most efficiently, yielding a minimum value for bsfc. This is not the case for two-cycle engines or for turbocharged engines.

In a *turbocharged engine,* where a turbine-driven compressor provides the intake under pressure, the torque and bsfc do not align. Figure 1.26 illustrates this. We analyze turbochargers in greater detail in Chapter 4. However, we can say that they deliver a greater mass of air to the cylinder, more fuel may be burned, and more power may be produced than in a self-aspirated four-stroke-cycle engine.

For two-stroke-cycle engines a set of performance curves will look like those in Figure 1.27. The torque curves of all diesels are relatively flat compared to those of spark-ignition engines. In this case there is a slight increase, then a decrease in torque with load. There is an increase in air supplied to the cylinder by the scavenging blower as the engine speed increases, resulting in the engine's ability to produce more work. As the speed increases, there is a decrease in the absolute time available for combustion, with a resultant decrease in torque. Thus the smoke limit may be reached even though the engine is supplied with adequate air.

Also, notice the *mechanical efficiency curve* included on this chart. What do you think it is? It is the friction horsepower (the difference between the indicated horsepower and the brake horsepower) divided by the brake horsepower. Since the mechanical efficiency is decreasing while the brake horsepower is increasing, it must mean that the indicated horsepower is increasing at a greater rate than the brake horsepower. It does. More fuel must be burned to gain each increment in horsepower

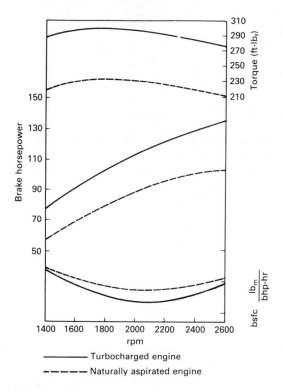

———— Turbocharged engine

– – – – – Naturally aspirated engine

Figure 1.26 Performance curves for a six-cylinder turbocharged (solid line) and naturally aspirated (dashed line) engine.

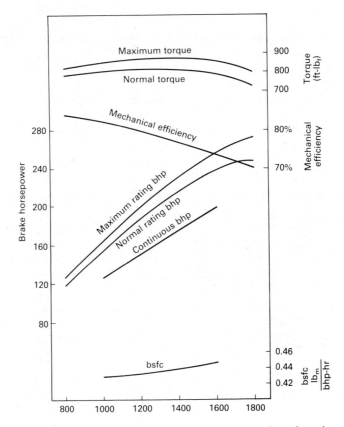

Figure 1.27   Performance curves for a six-cylinder two-stroke cycle engine.

than was burned for the previous horsepower rating. This means that the indicated work is greater—but the energy was not used productively. As the speed increases, the friction of the moving parts increases and the time for combustion decreases, so the percentage output is less. Also, the higher cylinder pressures cause higher bearing loads, hence more frictional energy.

CHAPTER

2

▮▮

Engine Components: The Parts and How They Function

In this chapter we examine in greater detail certain engine parts: the cylinder liner, the cylinder head, the crankshaft, the camshaft, the piston, and the valves. We will discover why the parts are formed as they are and areas of each part that may prove troublesome from an operating viewpoint.

CYLINDER LINER

The cylinder liner or sleeve is an insert in the cylinder block that protects the block from wear. The piston rings wear the replaceable liner rather than the block itself. There are two types of liners, a wet type and a dry type, the primary difference being that in the wet-type liner, the cylinder cooling water comes in direct contact with the liner, while in the dry-type liner, the water cools the block, which in turn, cools the liner. There is no direct water–liner contact.

Let's see what a liner looks like. Figure 2.1 illustrates a cutaway of a wet liner. The difference between this and a dry liner is slight—the water seal on the bottom and the raised portion just below the relief. Figure 2.2 examines the lower end of the liner when it is in the cylinder. There must be seals to prevent leakage of two liquids, water and oil. The water seal and oil seals do just this; the seals are usually neoprene O-rings, or a similar rubberlike material. The water circulates around the liner, removing heat that forms during the combustion process. Recalling from the first

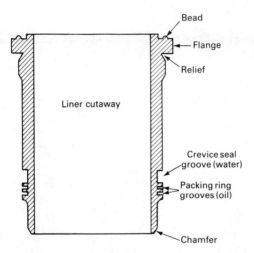

Figure 2.1 Liner nomenclature.

section that about 30% of the fuel's energy is dissipated as heat to the cooling water, the liner is the medium through which this heat passes to reach the water. Some heat flows directly from the combustion gases to the liner, while the rest flows through the piston, to the rings, and then to the liner.

In addition to being able to conduct heat away from the combustion chamber, the liner must be able to withstand the high pressures that occur during the combustion process. Special casting techniques are used and additionally the surface is treated to withstand the rapid piston movement without a high rate of wear.

Figure 2.3 illustrates the top end of the liner. The liner has a shoulder on it thats in the counterbored top of the cylinder. There must be this lip for the liner to rest on,

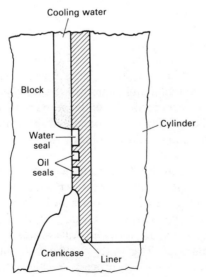

Figure 2.2 Illustration of liner oil and water seals.

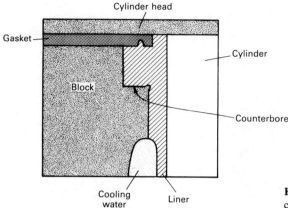

Figure 2.3 Illustration of the cylinder counterbore and head gasket.

or it would move, perhaps falling onto the crankshaft. The top of the liner must be sealed as well as the bottom. In this case a gasket is used. It forms a seal between the liner, block, and head. The head prevents the gasket from moving outward under high cylinder pressure as well as positioning the gasket. In addition, the gasket prevents water from leaking from the cooling passage in the cylinder or combustion gases from going in the opposite direction. You may have heard of a blown head gasket; this is it. Abnormally high cylinder pressure causes a break in the gasket; high-pressure combustion gases may then escape from the cylinder to the outside and/or into the cooling system. Neither is desirable.

Gaskets on diesel engines are deceptively simple in appearance, but there are many factors that must be considered in their design: peak pressure in the cylinder, combustion chamber design, liner type, all cast iron or bimetal engine forging, whether turbocharged, and where installed, factory or field. In addition, the gaskets provide a surface between such divergent parts as, for a head gasket, the liner, the head, cooling water, and combustion gases. It must be adaptable to the changes in size of different metal parts as their temperatures change, as well as being chemically inert with regard to the cooling system and the combustion gases. Gaskets are commonly made of several materials, often metal faced with a core of resilient material. As engine performance increases, the tolerances between surfaces is more critical; gaskets must not deform under load. For high-temperature gaskets, one material in common use has been asbestos. The use of alternative materials is increasing today because of the negative impact of asbestos on the environment.

A question may be nagging in the back of your mind; if the temperature in the combustion chamber is so hot, why doesn't the engine melt? After all, lead melts at 620°F, aluminum at 1200°F, cast iron at 2100°F, and steel at 2700°F. The reason is that the flame and the hot combustion gases do not have time to transfer enough heat to the cylinder walls (liners) and piston. Heat is transferred to these parts and they must be cooled or they fail—loose their strength and crack under the high temperature and pressure in the cylinder.

The average cylinder temperature is around 1150°F. Figure 2.4 illustrates a

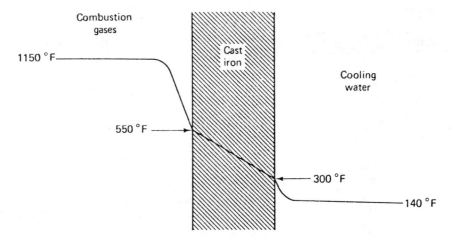

Figure 2.4   Typical temperature distribution across a cylinder wall.

temperature distribution across a cylinder wall. Cast iron, a typical liner and block material, loses its strength at temperatures above 550°F. We should note at this point that alloys can improve cast iron's strength; also, the average cylinder temperature can increase or decrease with load and rpm. Fortunately, a large temperature drop occurs at the gas–wall interface. This is because the gas molecules form an insulating layer (boundary layer) next to the metal surface. The same effect is true on the cooling-water side of the liner, but its effect is not as great because the density of water is much greater. Thus cooling with the resulting insulating layers provides thermal protection to the metal surfaces of the cylinder.

PISTON

We have seen how the cylinder is sealed on the top with the head, on the side with the liner, and now on the bottom with the piston. The piston is a movable seal. Figure 2.5 illustrates a trunk-type piston. Before going further, let's look again at Figure 1.9. Notice that the piston skirt, the lower part of the piston, will be forced to move slightly because of the connecting rod angularity as it moves around the crankshaft. The skirt must be able to withstand the side thrust forces developed by the angularity of the connecting rod movement. The side walls of the piston crown may be slightly tapered to allow for expansion due to high combustion temperatures. As the engine runs, the cylinder will become out of round because of the wear due to side thrust. The piston can no longer seal effectively, and the liner is removed, a new one installed, and the cylinder is "new" again.

The pistons are typically made of an aluminum alloy which conducts heat rapidly to the rings, and the rings transfer the heat to the liner. The rings have other purposes as well as heat conduction—sealing being an obvious one. Without the up-

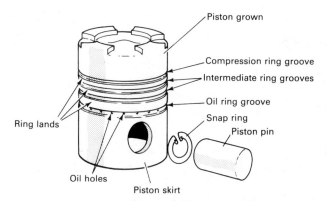

Figure 2.5   Piston nomenclature.

per compression rings, the pressure could not be raised as high on the compression stroke, and the combustion gases would blow by to the crankcase. The rings fit into ring grooves. What makes the rings seal so well? They have a certain spring to them, so the ring end gap is less when in the cylinder than when standing free. Figure 2.6 illustrates this. There must be some end gap clearance when the rings are on the piston in the cylinder to allow for thermal expansion. Figure 2.7 shows how the pressure of the gases in the cylinder cause the ring to move against the liner, sealing the cylinder. If carbon builds up behind the rings, the rings stick and will not seal properly.

The lower rings are called oil control rings. Their purpose is to distribute oil in a uniform film over the cylinder wall and to prevent oil from being drawn into the combustion chamber when the piston is on the intake stroke and a vacuum exists in the cylinder. How is this done? Figure 2.8 illustrates oil control ring action on the downward and upward strokes. On the downward stroke the ring scrapes the excess oil from the surface. The oil drains back to the sump through drain holes in the ring lands. Without this draining ability, the rings could not remove all the oil. The oil is evenly distributed by the same rings on the upward stroke. The oil also removes some heat from the cylinder wall in this process. Of course, the pressure with which the oil control rings push against the cylinder wall determines how completely the oil is removed. It should not all be removed, as the compression rings need lubrication;

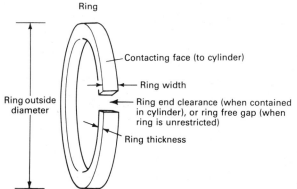

Figure 2.6   Piston ring nomenclature.

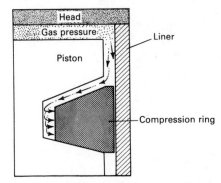

Figure 2.7 Illustration of gas pressure causing the ring to seal against the liner and piston.

otherwise, scoring and scruffing of the liner surface occurs. The oil rings have some type of metal expander which forces them snuggly against the wall. This is determined by the manufacturer.

A ring pack is the arrangement of compression and oil control rings that a piston has. The type of rings and the arrangement varies as the engine duty requirements vary. For instance, a naturally aspirated engine will have one arrangement, but it is different from that in a turbocharged engine. The type of turbocharging will also affect the ring pack.

Engine loading is one of the key factors in compression ring selection, of course, but it is not simply the amount of the loading but also the rate of increase in that loading from zero cylinder pressure to peak cylinder pressure. A higher brake mean effective pressure typically dictates a move from a heat-treated cast iron compression ring to a ductile iron ring. Increasingly higher modulus of elasticity and bending strength are generally called for, as diesel engines are required to have greater performance.

Referring again to Figure 2.5, we note that there is one more part associated with the piston—the piston pin. The piston pin must transmit the force from the piston to the connecting rod on the downward stroke, and vice versa on the upward

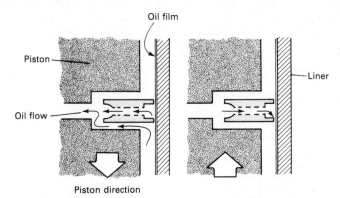

Figure 2.8 Oil control ring removing oil from (downward stroke) and distributing oil to (upward stroke) the liner.

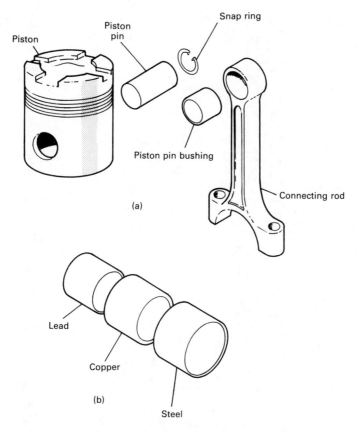

Figure 2.9   (a) Location and (b) schematic of a piston pin bushing.

stroke. Figure 2.9a illustrates how the piston pin and connecting rod fit together. The forces on the piston pin bearing, a bushing in this case, are severe. The bushing wears, and can be replaced. It is much less expensive than replacing the entire connecting rod. The piston pin is held within the piston by a snap ring, preventing any side movement of the piston pin. Figure 2.9b illustrates a bushing. Bushings are a composite of three materials: a steel backing, a copper lining for long wear, and a thin lead-type inner shell which conforms to surface irregularities of the bearing surface.

The connecting rod transforms the reciprocating motion of the piston to the rotary motion of the crankshaft. As we have just seen, it is attached to the piston via the piston pin. The connecting rod must be able to withstand the large forces that are created by the piston motion and do so while at any angle, requiring greater strength. The connecting rod is drilled throughout its center (Figure 2.10). Lubricating oil, under pressure, flows through this passage to the piston pin bushing, assuring an adequate supply of oil, which prevents the metal surfaces from touching.

The lower end of the connecting rod is attached to the crankshaft. The crankshaft must be able to rotate, so a bearing is used. Note that the connecting rod has

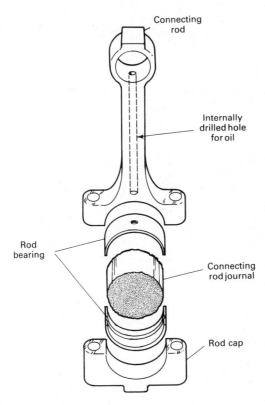

Figure 2.10   Connecting rod.

two parts which, when bolted together, hold the bearing firmly in place. A sketch of a bearing is shown in Figure 2.11. It has similar material to that of a bushing. As the engine runs there will be normal wearing of the bearing due to friction. The lubricating oil film minimizes the wear but cannot stop it.

What is this oil film? Consider Figure 2.12a, which illustrates a shafting resting in a bearing. Oil surrounds the shaft where it can, but at rest there will be some metal-to-metal contact. As the shaft turns, oil is pulled between the metal surfaces and there is no contact, as shown in Figure 2.12b. The pressure of the oil in the wedge below the

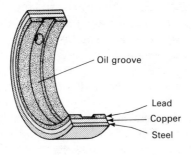

Figure 2.11   Schematic of a bearing.

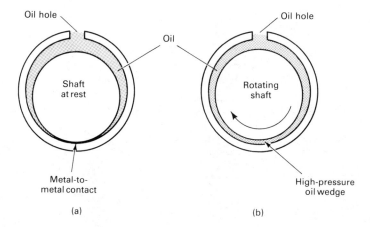

Figure 2.12 Principle of an oil wedge.

shaft may reach several hundred psi. When the shaft stops, the oil begins to be squeezed out by the weight of the shaft. There will be partial lubrication when the metal surfaces contact as the oil fills the microscopic valleys on the metal's surfaces. The longer the metal surfaces lie on one another, the greater will be the metal-to-metal contact. Thus the wear of metal on starting will be greater. This is why frequent starting and stopping increases wear.

In the case of the connecting rod bearing, the upper bearing wears more quickly than the lower one. This is because the power stroke of the piston causes higher pressures—forces—on this half. It also causes the upper half of the piston pin bushing to wear more quickly, for the same reason. Bushing and bearing wear will cause the volume at TDC to increase and hence the compression ratio to decrease. There is a loss of performance because of this. Also, if the wear is too great, the oil film cannot support the loads, and scoring and rubbing of the metal surfaces will occur.

CRANKSHAFT AND CAMSHAFT

Figure 2.13 illustrates a crankshaft for a six-cylinder engine. This rests in supports, bearing cradles, forged and then machined in the cylinder block. On the top of the block there are the forces due to the combustion process, and on the bottom the transformation of these thermal forces must be withstood in the form of the mechanical force exerted on the crankshaft by the connecting rod. One-half of a bearing shell rests in these cradles. The crankshaft main bearing journals rest in the bearing shells. These are denoted by the letter "M" in the sketch. Notice that they are all in a line. The surfaces of the journals are carefully machined, so there is a very close fit between the bearing and the journal, which are separated only by a thin film of lubricating oil.

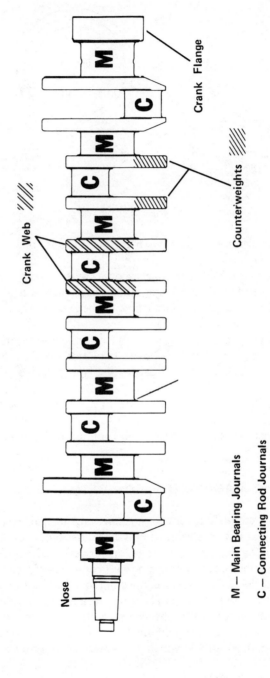

Crank Flange

Counterweights ⦀

Crank Web ⦀

Nose

M — Main Bearing Journals

C — Connecting Rod Journals

Figure 2.13 Schematic diagram of a crankshaft. (Reproduced with permission from Cummins Engine Co., Inc.)

The crankshaft is drop-forged. Forging is the process of shaping metal, typically red-hot metal. Today, rather than a blacksmith beating the iron into shape, hammers force the metal into dies (molds) to achieve the desired shape. A drop hammer is used in drop forging. The journal surfaces are specially hardened and then machined to remove any surfaces that could increase the metal's stress. The bearing cap holds the crankshaft in position and houses the other half of the bearing shell. Passages are drilled through the crank web to admit oil from the main bearing to the connecting rod bearing and then through the connecting rod. The connecting rod journals, denoted by the letter "C" in the sketch, are where the lower end of the connecting rod attaches itself to the crankshaft.

There are seven main bearings and six connecting rod bearings in a six-cylinder engine. The engine would have a firing order of 1-5-3-6-2-4, cylinder 1 being the one nearest the front of the vehicle. It is desirable that the same number of pistons firing during each revolution of the crankshaft, in this case every 120 degrees, which allows a balance of forces in the engine. This means that there are always two pistons traveling in the same plane at the same time. For instance, consider pistons 1 and 6. Number 1 may be at the end of the compression stroke and number 6 at the end of the exhaust stroke. They are 360 degrees out of phase with each other but traveling in the same direction. It is also desirable that pistons next to each other do not fire sequentially or simultaneously. This would create a high heat load on the metal surfaces, which could cause metal fatigue. Also, the lubricating oil film could be vaporized by the higher-than-normal temperatures. If we consider these parameters as being, first, that the pistons be 120 degrees apart in firing; second, that adjacent pistons do not fire together or sequentially; and third, that cylinder 1 usually fires first—then a logical firing sequence is 1–5–3–6–2–4. Notice that the connecting rod journals indicate that the pistons are 360 degrees out of phase.

We noted that the engine cylinders fired as evenly as possible to minimize an imbalance of forces on the engine and on the crankshaft. This does not eliminate the imbalance, however. Three additional devices are used in this regard. The first are counterweights. Because the crankshaft rotates and there are parts, such as the connecting rod journals and the crank webs, which are off-center, an unbalanced centrifugal force will act on the crankshaft, causing it to vibrate. To balance these forces, counterweights are used. Although it may seem that these might make matters worse, they balance the natural centrifugal force on the crankshaft with one in the opposite direction and equal in magnitude. The second device is a flywheel, not shown, but attached to the crankshaft flange. The diesel is a reciprocating engine, which means that the power production is pulsing. The flywheel has a large mass and rotates at engine speed. It stores energy by this rotation. It smoothes out the power strokes of the engine. The flywheel has two nonvibration purposes: It is the mounting surface for the clutch and it acts as the clutch's friction surface and its outer rim has a ring gear shrunk-fit onto it. This gear is used by the starting motor. The third device is a vibration damper, which is attached to the front (nose) of the crankshaft. Because the initiation of the downward power stroke forces the crankshaft out of alignment, then

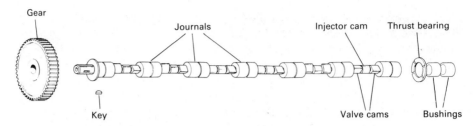

Figure 2.14 Schematic diagram of a camshaft. (Reproduced with permission from Cummins Engine Co., Inc.)

moves it quickly back into alignment, the oscillation could cause crankshaft failure. This is resisted by vibration dampers, which slow the rate of change of the crankshaft.

Also attached to the crankshaft at the front end is a gear. This gear, by means of a chain or using intermediary gears, drives the camshaft, shown in Figure 2.14. The gear on the front of the camshaft is twice as big as that on the crankshaft for four-cycle engines. This is because the camshaft turns at one-half the crankshaft speed. The camshaft is supported by bearings which act on its journals. There are three cams or lobes for each cylinder: one for the intake valve, one for the exhaust valve and one for the fuel injector. On two-stroke-cycle engines the intake valve does not exist, but exhaust and fuel injection are required. The camshaft is supported by a bearing journal between each cylinder. On some engines there may be two camshafts, one for the intake valves, one for the exhaust valves. Also, not all engines have the fuel injector activated by a cam on the camshaft.

Figure 2.15 illustrates two types of cam followers. In both cases pushrods force the rocker arms to open the valves or injector. The followers maintain contact with the cam surface. Refer to Figure 1.4 for the diagram of the cam. There must be clearance between the rocker arm and the valve or injector when the follower is on the base circle of the cam. As the temperature of the valve increases, the valve will expand and should there be no clearance between the valve and the rocker arm assembly, the valve would remain open. That would allow a constant flow through the valve, which cannot be allowed. If the clearance between the rocker arm is too large, the valve will open late and close early. Valves are timed when the cam follower is on the base circle of the cam. A feeler gauge, calibrated strips of metal, are used to determine the clearance between the rocker arm and the valve. Adjustments may be made to the rocker arm to vary the clearance. In Figure 1.4, "ramp" denotes the distance required, moving on the cam flanks, to remove the valve clearance. The shorter the ramp, the less distance or time it takes for the clearance to be removed between the valve and the rocker arm. The faster the change, the greater will be the force on the flank and the greater will be the wear.

Figure 2.16 illustrates a dual overhead cam. Notice that the cam acts directly on the cam follower (the roller), which is attached to the rocker arm. The rocker arm acts on the valve. The pushrod is eliminated in this case. Pushrods must have the strength

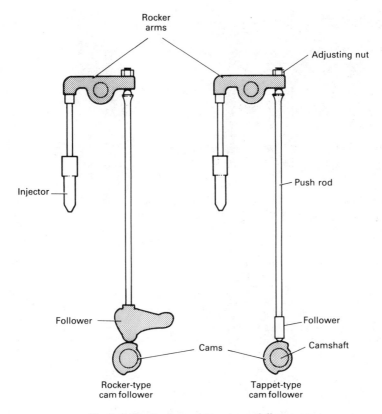

Figure 2.15   Rocker and tappet cam followers.

to transmit the force required to open the valves and injectors. Some pushrods, called push tubes, are hollow, to reduce their mass, but still meet the strength requirements. Because their weight is less, their inertia is less. Inertia may be thought of as the desire not to change. The inertia of the parts affects the camshaft in two ways: in causing the pushrod and rocker arm (valve gear) to move up, and creating an impact with the valve gear as it moves down. In high-speed engines the pushrod may want to keep moving up, due to acceleration, even though the nose of the cam has been passed and it should readjust and move down. This can cause the follower to leave the cam surface ever so slightly. If it does, the valve gear catches up with a bang, pounding back onto the surface, which is obviously not good for the cam face. The valve spring should have enough force in the opposite direction to prevent this.

The last part of the camshaft we will discuss is the thrust bearing. What is it used for? The interaction between the cams and the cam follower produces a slight horizontal force. This is because the surfaces between the cam and the cam follower are not perfectly parallel; there is a slight angle. This means that while almost all the force is directed toward opening the valve, a small component of force tends to move the camshaft. The thrust bearing prevents this sideways movement.

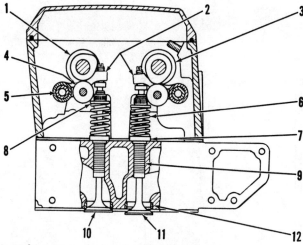

1. Exhaust camshaft
2. Rocker-arm assemblies
3. Inlet camshaft
4. Roller
5. Rocker-arm shaft
6. Valve spring
7. Valve rotator
8. Valve retainer and lock
9. Valve guide
10. Exhaust valve
11. Inlet valve
12. Valve-seat insert

Figure 2.16   Valve mechanisms.

VALVES

Figure 2.17 illustrates a typical valve and shows its terminology. "Valve face" refers to the contact area on the valve and valve seat in the cylinder head. The seat is the mating part on the cylinder head that the valve rests against. The valve seat is angled at either 45 or 30 degrees from the horizontal. The angle is determined by a variety of factors, such as the fillet curvature, the valve lift, and whether the engine is naturally

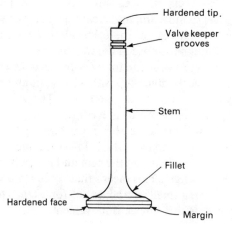

Hardened tip.

Valve keeper grooves

Stem

Fillet

Hardened face

Margin

Figure 2.17.   Valve nomenclature.

aspirated or supercharged. The 30-degree valve has less restriction across the seat, allowing the flow of gas (air or exhaust) to start earlier and end later. The 45-degree seat has a greater seating force, less chance of leakage, and a greater velocity across the seat, cleaning it of carbon buildup. Because the seating force is greater, the valve seat has a tendency to deform. The intake valve is about 40% larger than the exhaust valve, allowing an easier, hence greater, flow of air into the cylinder.

In Figure 2.16 we see how the valves are held in the head. The valves are actuated by the rocker arms, which contribute some angularity to the up-and-down motion of the valve. The angularity would cause the valve stem to wear against the cylinder head, and eventually the head would wear and combustion gas could escape through the stem–head clearance. Just as liners are used in the cylinder to prevent the block from wearing due to piston motion, valve guides are used to prevent head wear due to valve motion. Valve guides are shown in Figure 2.16.

If there is excessive wear, measured as the clearance between the valve stem and the valve guide, on the exhaust valves, there may be carbon formation in the valve guides, as exhaust gases will collect in that area. Carbon forms and the valve stem becomes sticky and tends not to move—to hang up. Also, as the valves become sticky from carbon buildup, they also tend to overheat, which worsens the problem of valve deterioration. Excessive intake valve guide wear also causes high engine lubricating oil consumption because the oil from the rocker arm assembly is pulled through the guides on the piston's intake stroke.

The transfer of heat from the valves to the head occurs through the valve seat. The seats must be smooth, not only to prevent leakage but also to provide for good heat transfer. Pits on valve seats must be removed by a grinding compound, or if the seat is very scored, they must be refaced. Normal wear, over a long period of time, will require that the valve seats in the head be refaced. Should this be done more than two or three times, the head valve seat diameter increases significantly, causing the valve to seat deeper in the head and restrict the gas (air, exhaust) flow. A new head is then in order.

Some engines have valve seat inserts incorporated into the head, which wear instead of the cylinder head. Figure 2.18 illustrates these, shown as solid black areas. These are replaceable. As the inserts are ground away and exceed the dimensions required, new ones may be inserted.

There are also limitations on grinding the valve seats on the valve, which also pit

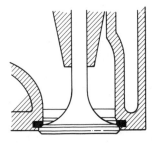

Figure 2.18 Valve seat insert (shown as solid black areas).

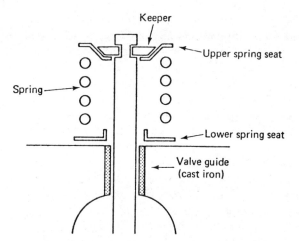

Figure 2.19   Valve guides.

and score. One result of excessive grinding is the creation of too sharp an edge at the valve seat. This is created as the grinding causes the valve seat to incorporate material in the valve margin. Excessive temperature buildup in this high seat stress area can cause the metal to fatigue and droop, a problem called *tuliping,* as the valve resembles a tulip.

Figure 2.19 illustrates a simplified view of the valve retainer that can be seen in context with the other parts of the valve in Figure 2.16. The upper spring seat is held in place by a keeper. The valve spring must be compressed to get the keeper on, as it is split in half. The spring pressure holds the valve shut and the upper spring seat tight. The lower spring seat prevents the spring from pushing directly on the head. The last feature that is used with valves is a valve rotator. Each revolution of the engine actuates the rotator, which rotates the valve slightly. This minimizes uneven wear of the valve seats.

The lower surface of the cylinder block is open as the crankshaft rotates here. A flange is located around the periphery of the block, and attached to this is an oil pan which contains the engine lubricating oil. A gasket is used between the pan and the block to seal any irregularities on the metal surfaces. The volume enclosed by the oil pan is called the *crankcase,* as it has the crankshaft within it. The lowest part of the oil pan is called the *sump.*

3

The Fuel System

The fuel system supplies the energy to the engine in the correct proportion to meet the load requirements on the engine. The fuel must be correctly metered, timed, and injected and the engine must be controlled, preventing overspeeding or stalling. This system is the most complex engine system and the one with the least commonality between engines. Before we get to the components that may vary from engine to engine, let's discover what diesel fuel is and how it is rated.

FUEL CHARACTERISTICS

Diesel engines use a refined crude petroleum which has had some of its more volatile constituents removed and does not include the heavy residual fuels. There are, of course, different grades of diesel oil, so let us start at the beginning, with the crude petroleum. The crude petroleum, or crude oil, is that found in a natural state beneath the earth's surface. It was formed over long periods of time by decayed organic matter, subjected to high temperatures and pressures found within the earth's crust.

The composition of the crude varies from place to place as a function, in part, of the matter forming it. The primary components are carbon, hydrogen, and sulfur, with traces of many other elements. Carbon normally accounts for 82 to 87%, hydrogen 10 to 14%, and sulfur (and chemical trace elements) 1 to 6%. The heating value, or chemical energy of the fuel, is a function of its chemical composition; the greater the percentage of hydrogen, the greater the heating value.

To understand more completely how the oil gets from the ground to the diesel engine, let us examine some of the simpler methods of distilling the fuel. The crude oil contains many different hydrocarbons, some of which are more valuable than others. By distilling the fuel, the more valuable (typically, more volatile) components are evaporated and then collected. The process continues until only very heavy oil residue remains. Elementary distillation processes rely on the difference in boiling points of the various hydrocarbons. When using such processes, 20% of the crude oil could be converted into gas or gasoline, 40% into distillate fuel and kerosene (diesel fuel is this group), and 40% into residual fuel. For economic reasons, it is preferable to have less residual and distillate fuel and more gas or gasoline. To this end, modifications of the straight-run distillation process have been devised. By introducing thermal crack-ing—breaking the hydrocarbon molecule into lighter and heavier hydrocarbon molecules by subjecting the crude oil to high temperatures and pressures—the distribution improves to 35% gas or gasoline, 35% distillate, and 30% residual. Finally, at least for our introductory view, catalytic cracking, or catalytic reforming, may be considered as well. In the catalytic cracking process, a catalyst promotes the breakdown of residual hydrocarbons by high temperature and pressure. The catalyst does not enter the reaction, but it will absorb certain molecules. In this case, the distribution rises to 48% gas/gasoline, 43% distillate, 5% residual, and 2% coke. Coke is a carbon residue left by the cracking processes and is used in furnaces as a fuel.

The fuels that are used in high-speed diesel engines are graded 1D and 2D. For some medium-speed engines a grade of 4D is used. There are a variety of factors which go into the grading of the fuel. Before we examine these, 1D fuel is used in engines requiring frequent changes in speed and load and 2D fuel is used in industrial constant-speed engines and in some large variable-speed engines.

Whereas the octane number defines the quality of gasoline, the *cetane number* defines the ignition quality of diesel fuel. We remember that all diesel fuels have igni-tion delay, the time between their injection into the combustion chamber and when combustion first starts. A certain hydrocarbon called cetane is given a rating of 100 when tested under certain engine conditions. A blended fuel, such as 1D, is tested under the same test conditions and has a certain ignition delay. A blend of cetane and another hydrocarbon, alpha-methylnaphthalene, which has a zero cetane rating, are mixed until the blend has the same ignition delay as the test fuel. The percentage of cetane in the blend is called the cetane number of the fuel. For instance in the case of 1D, the cetane number may be 45, which means that the blend was 45% cetane and 55% alpha-methylnaphthalene. The higher the cetane rating, the shorter the ignition delay. Engines run at high speed require a higher cetane number than those run at a lower speed because the time for combustion is very small and extended ignition delay would mean that the fuel would not have time to completely oxidize.

Table 3.1 lists typical values for 1D and 2D fuel as suggested by ASTM (American Society for Testing and Materials). We will discuss these briefly, as often you have no control over the fuel you can buy. However, it is wise to be aware of the factors that go into designating fuel quality.

Specific gravity is the ratio of the density of a substance to the density of water, both measured at 60°F. This helps us relate the volume and mass of the substance to that of water.

API gravity is an arbitrary type of specific gravity defined by the American Petroleum Institute (API). As the specific gravity decreases, the substance weighs less than water per unit volume; its API gravity increases. In general, as the API gravity increases, the cetane number increases.

The *Btu per gallon* value is the product of the fuel's heating value per pound times the number of pounds in a gallon. Actually, the 1D fuel has a greater heating

Table 3.1 Selected Diesel Fuel Properties

ASTM Grade	Specific Gravity (Average)	API Gravity	Btu per Gallon	Distillation Temperature (90% Point)		Viscosity at 100°F (centistokes)		Pour Point	Flash Point	Ash Content (% wt)	Carbon Residue (%)	Cetane Number Minimum
				Minimum	Maximum	Minimum	Maximum					
1D	0.834	35–40	137,000	—	550°F	1.4	2.5	10°F below ambient	100°F or legal requirement	0.01	0.15	40
2D	0.877	26–34	141,800	540°F	640°F	2.0	4.3	10°F below ambient	125°F or legal requirement	0.02	0.35	40

value (more hydrogen atoms) but contains fewer pounds per gallon than does 2D fuel, which results in its having less total Btu in the gallon.

The *distillation temperature* is another means of distinguishing fuel quality. We saw that the lower-molecular-weight hydrocarbons vaporized at lower temperatures than the higher-molecular-weight hydrocarbons which are part of diesel fuel. There will be an increasing series of distillation temperatures, at which more and more of the fuel is vaporized. The temperature is recorded for each 10% of fuel collected, with the 90% point and the end point being specified. From Table 3.1 we see that for 1D fuel the maximum 90% point is 550°F, while for 2D it is 640°F. The higher the temperature, the longer it takes to vaporize the fuel in the combustion chamber of an engine. For high-speed engines 1D is preferred because of its greater volatility, even though the cetane number is the same.

Another important fuel property is *viscosity*. It is experimentally determined by measuring the time it takes for a specified amount of liquid to flow through an orifice of a certain size. Figure 3.1 is a schematic of how oil viscosity is determined. The oil bath is used to warm and maintain the sample at the desired temperature, in this case 100°F. We also see that there is a significant difference between 1D and 2D viscosity. The fuel's viscosity affects the combustion process, with a high-viscosity fuel having coarser spray droplets and penetrating farther into the combustion chamber. The larger the droplets, the longer it takes to vaporize them during the combustion process. If the viscosity is too low, the fuel may be too volatile and may not penetrate sufficiently into the combustion chamber to assure good air–fuel mixing. Additionally, the fuel oil is used as a lubricant for certain parts, and a low viscosity reduces the fuel's lubricating value.

The *pour point* refers to the minimum temperature at which the fuel can flow through the filters and reach the fuel injection pump. It is specified as being at least

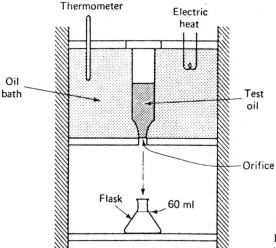

Figure 3.1 Schematic diagram of a Saybolt Universal Viscosimeter.

10°F below the lowest ambient temperature. Problems arise, of course, if the ambient temperature is below the pour point.

The *flash point* is the temperature to which the fuel must be heated to allow flammable vapors to form on its surface. These vapors can be ignited by an open flame held above the surface. The importance of the flash point is in storage safety, not engine performance.

The *ash content* of oil is that material remaining after the fuel has been burned at a high temperature with only incombustibles left. Ash will accelerate engine wear through abrasion.

The *carbon residue* is the solid carbon left after the fuel is burned with a fixed amount of oxygen. This correlates with the carbon-forming characteristics of the fuel, which cause such problems as ring sticking and deposits with the combustion chamber.

At this point let's reflect a moment. There are many characteristics of a fuel, few of which we as purchasers have control over. There are significant differences between 1D and 2D fuel, and using 2D fuel where 1D is specified by the engine manufacturer may result in accelerated engine wear and poor engine performance. Factors such as engine speed and engine design must be considered before making such a change.

OVERVIEW OF THE FUEL SYSTEM

The fuel system has two distinct purposes: supplying the fuel to the engine via tanks, lines, filters, and pumps; and controlling the flow of fuel into the cylinder. The control part of the fuel system will vary with engine type, so the manufacturer will have more than one type of control system, considering the various operational requirements of the engine. The supply system is reasonably standard, in that the same components are in all systems.

Figure 3.2 illustrates a typical supply system for an engine using unit injectors. These will be discussed shortly. Usually, there are at least two filters in the fuel system, one before the fuel transfer pump and one after the fuel transfer pump. The purity of the fuel is very important to injector operation, and every effort is made to assure that the fuel entering the injector contains no particles of metal or dirt. The fuel transfer pump is a gear-type pump, which means that it can pull oil into it from the tank as well as discharge the oil from it at a higher pressure. The pressure regulators in the system create a resistance to the flow of fuel so that there is a pressure buildup of fuel in the fuel pump. There has to be a pressure maintained on the return line from the injectors or the pressure in the supply line would drop drastically. The combination of the two pressure regulators assures that the fuel will be supplied under pressure to the unit injectors. We also see that a hand priming pump is located on the fuel system. Its purpose is to remove air from the system when the system has been disassembled for repair. Air is the enemy of any hydraulic system, as the air is compressible and occupies a variable volume dependent on the

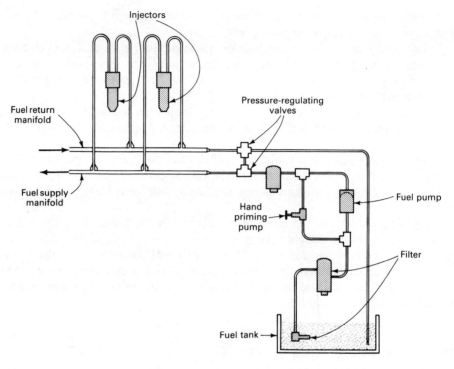

Figure 3.2 Unit injector fuel system.

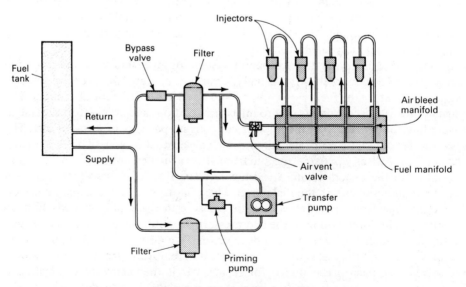

Figure 3.3 Nonreturn fuel system.

pressure and will not allow the flow of oil in parts of the system where it becomes entrapped.

There are other fuel supply systems which do not have a return line from the injectors. Figure 3.3 illustrates such a system. The fuel transfer pump delivers the fuel to a fuel injector pump, which meters and times the correct amount of fuel to the various injectors. In this case each injector has its own fuel injector pump, separate from the injector itself.

We are now at the crossover point between supply and control. Fuel injectors and fuel injector pumps control the amount of fuel that goes into the cylinder, but not independently. The engine governor adjusts the pump or injector setting so that the fuel supplied to the cylinder meets the engine's needs. We will continue with the equipment in the fuel system before analyzing how the engine governor operates.

FUEL PUMP

The fuel pump in high-pressure fuel systems increases the fuel's pressure to several thousand psi and delivers this fuel through high-pressure fuel lines to the injectors. In this type of system, there may be a pump for each injector, a multipump system, or one pump supplying all injectors, called a distributor pump system.

In the *multipump system* the force for pressurizing the fuel comes from the camshaft. The fuel pump plunger rides on the camshaft, via a cam follower. The pump plunger is always lifted by the cam and makes a full stroke. The fuel is compressed by the plunger and this high-pressure fuel is sent to the injector. How does the correct amount of fuel get sent to the injector? There must be a metering aspect to this pumping action. The key to the operation is the rotary plunger, illustrated in Figure 3.4. Notice that a helix is formed on the pump plunger and that the left- and right-hand sides have ports which are opened to the fuel supply coming from the fuel transfer pump, or supply pump. When the plunger is below the ports, fuel from the transfer pump is everywhere throughout the system. It is below and above the helix area. A very close fit is required between the barrel and plunger, as the fuel pressure may reach 3500 psi during the injection process and there can be no leakage. As the plunger moves vertically upward, the ports are closed and compression begins. The oil from the pump to the injector increases in pressure. Very soon the pressure is high enough to open the injectors and injection begins. No more fuel can enter the barrel once the plunger covers the ports. As the plunger continues upward, fuel is forced through the injector and into the cylinder. As the helix moves up and uncovers the port, the fuel flows from the high-pressure region to the low-pressure supply region. This is called the *end of injection*. Thus the pump has a constant beginning and a variable ending, depending on where the helix uncovers the port.

Fast plunger acceleration during the beginning of injection is desirable so that the pressure builds up quickly, allowing the injection to open quickly and cleanly, preventing the dribbling of fuel from the injector tip. Also, a slow pressure increase delays the time when the fuel is injected into the cylinder. The conclusion of injection should also stop quickly, to prevent fuel from tapering off and dribbling at the injec-

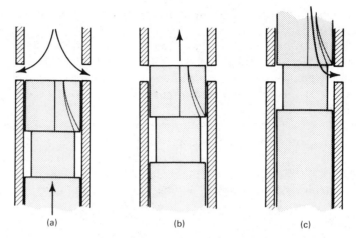

(a)   (b)   (c)

Figure 3.4 Injection sequence for a rotary plunger pump: (a) ports open, being closed; (b) beginning of injection; (c) end of injection.

tor tip. The shape of the fuel cam is designed to control acceleration and deceleration of the pump plunger. A potential malfunction due to dribbling of fuel at the injector tip is *coking,* the formation of carbon deposits in the nozzle. When this occurs, the fuel does not burn completely and solidifies in small nozzle passages, forming cones on the outer surface of the injector tip. The carbon formation distorts the spray pattern of the nozzle, causing improper combustion, possibly black smoke, and possible fuel impingement on the piston crown because of poor atomization.

We see that by rotating the plunger, the fuel supplied to the injector will vary, because the ending varies. The engine governor controls the plunger rotation. As engine load demands more or less fuel, the governor moves to supply it by rotating the plunger. The barrel remains stationary.

How does the engine stop? The fuel must be cut off. Figure 3.5 illustrates this. In this case there is an opening between the top of the plunger and the inlet port, which prevents fuel from being pressurized. In this sketch we see how the effective stroke of the plunger, when fuel is being pressurized, varies with plunger rotation, even though the actual plunger movement is the same in all cases.

The high-pressure fuel lines connecting the fuel pump with the injector are of a fixed length and diameter which should not be altered. Let us see why. Assume that the pressure required to open the injector is 3500 psi, so the fuel pump will compress the fuel until the pressure reaches 3500 psi. Once the injector opens, less pressure is required to keep it open, say 3200 psi. This allows an injector to snap open, as we shall see. Because of this, a pulsating pressure shock wave moves back and forth between the injector and the fuel pump in the high-pressure fuel line. Let this have a value of 400 psi. Assume that injection ends and the injector closes because the pressure is slightly less than 3200 psi. When the reflected shock wave comes back to the injector the pressure is 3600 psi, the injector snaps open, and combustion continues at a time when it should not. This is called *after* or *secondary combustion.* To

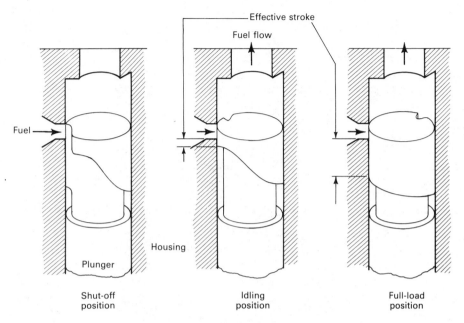

Fuel

Housing

Plunger

Effective stroke

Fuel flow

Shut-off
position

Idling
position

Full-load
position

Figure 3.5 Fuel pump plunger operation.

prevent this, a delivery valve or discharge valve is installed in the fuel pump. The delivery valve has a retraction piston which compensates for the expansion of fuel oil by occupying a certain volume in the high-pressure line between the pump and the injector. When the fuel pump port opens, allowing the fuel oil pressure to drop, the retraction piston volume is available for fuel, causing a rapid decrease in the fuel pressure in the high-pressure fuel line. Thus even with the 400-psi shock wave traveling in the high-pressure fuel line, the fuel's pressure drops significantly below 3200 psi, preventing the injector from opening due to shock wave pressure. The size of the retraction piston is determined separately for each system, as it depends on the total volume of fuel in the high-pressure fuel line. Fuel has a compressibility of 1% at 3000 psi and the retraction piston must account for this. Thus the supply line volume to the injector is critical in preventing after-injection. Replacement lines should always be of the same length and diameter.

Not all systems have a separate fuel pump and injector. A *unit injector* combines the features of both. There is a constant fuel oil supply provided to the unit injector which meters, compresses, and injects the oil.

Before we leave the topic of fuel pumps, it is possible that the helix on the pump plunger be on the top, rather than the bottom as previously illustrated. Figure 3.6a illustrates this case. In this case the beginning of injection is not constant, as before, but variable. The rotation of the plunger allows compression, hence injection, to occur earlier or later during the plunger movement. Figure 3.6b illustrates a plunger that is variable beginning, variable ending. Why have a variable beginning? The variable beginning allows the governor to advance injection as speed increases. Other

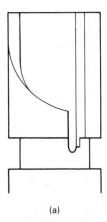 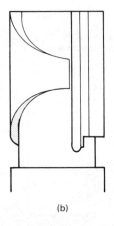

Figure 3.6   Plunger of (a) upper helix design and (b) upper and lower helix design.

(a)   (b)

designs may use a separate advance mechanism. Also, there may be different types of fuel metering pumps in use, but 90% or more are of the rotary plunger variety.

Cummins engines use a mechanical injection system in which the injector is opened by the action of the fuel cam. It forces the injector plunger down, and atomized fuel enters the combustion chamber. The system for providing the correct amount of fuel is called a pressure–time (P–T) system, because the fuel metered to each injector is a function of pressure and time. The governor causes a gear-type fuel pump to change pressure; the greater the pressure, the greater the amount of fuel that enters the injector. The time part of the system is the time the fuel has to enter the injector body before injection. These factors allow the fuel to vary with engine demand. Additionally, the Cummins air fuel control (AFC) valve limits the fuel to the injectors to be compatible with air from the turbocharger during acceleration, preventing incomplete combustion and a smokey exhaust. This control device operates only during this transient condition.

INJECTOR

There are two types of fuel injectors in common use: the hydraulic injector, used with the high-pressure fuel pump, and the unit injector, used in conjunction with the low-pressure fuel pump.

Let us consider the *hydraulic injector* first. Figure 3.7 illustrates the fuel injector. There are three main parts to the injector: the nozzle, the nozzle body and the cap nut. Figure 3.8 gives a detailed view of the nozzle assembly. If we follow the flow of fuel as it leaves the metering pump and enters the injector, we notice that it proceeds through the inlet adapter and through a drilled passage in the nozzle holder to the nozzle itself. This part of the nozzle is the pressure gallery near the nozzle tip. This is the area below the tapered part of the needle valve. The fuel pressure is resisted by the spring, which is directly connected through the spindle. When sufficient opening pressure for the nozzle is reached, the needle valve snaps off its seat, allowing the flow of fuel through the orifices at the nozzle tip, and injection begins. The reason the

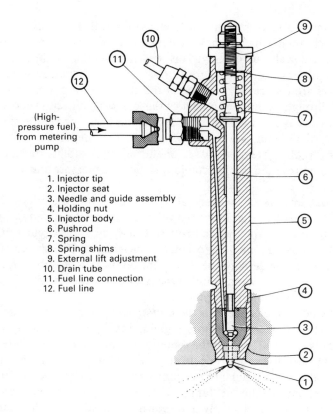

1. Injector tip
2. Injector seat
3. Needle and guide assembly
4. Holding nut
5. Injector body
6. Pushrod
7. Spring
8. Spring shims
9. External lift adjustment
10. Drain tube
11. Fuel line connection
12. Fuel line

(High-pressure fuel) from metering pump

Figure 3.7 Multihole injector assembly.

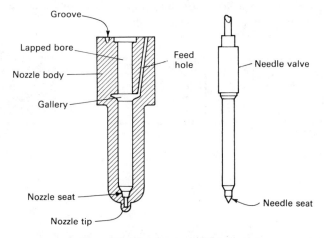

Groove

Lapped bore

Nozzle body

Gallery

Feed hole

Needle valve

Nozzle seat

Nozzle tip

Needle seat

Figure 3.8 Nozzle nomenclature.

needle valve snaps off its seat is that the fuel, as the needle first rises, is able to act on a greater surface area, which almost instantaneously overcomes the spring force, pushing the needle valve up quickly. This prevents dribbling at the start of injection.

The spindle and needle valve are lubricated by the slight flow of fuel past them. The tolerances are very close, to allow for the quick pressure rise. Additionally, this flow of fuel through the nozzle and out of the leak-off connection cools the injector. The leak-off valve is also used when the system has been opened for repair and there is a possibility of air in the fuel lines. The fuel is bled through the injector, eliminating the air. All hydraulic systems are based on the incompressibility of fluids. The inclusion of air would result in unpredictable variations in pressure due to the collapse of air bubbles.

The nozzle may be of two general types, multiorifice and pintle. Figure 3.9 illustrates these. The tip of the pintle need valve has a diameter slightly smaller than the hole in the injector tip. The fuel, in being forced through the circular orifice formed by the pintle nozzle and the hole, becomes cone-shaped. The choice of nozzle type becomes one of engine design—the use of precombustion chambers for instance, as mentioned in Chapter 1. Some pintle-type nozzles open outward; the needle valve moves downward, out into the combustion space.

The second type of injector is the *unit injector,* illustrated in Figure 3.10. It is often associated with General Motors engines, although it is certainly not limited to them. Typically, it is clamped on the center of the cylinder head. There are two fuel lines to the injector: the supply line through which the fuel enters the injector body, flowing through the filter and into the body itself; and the return line from the injector, which goes back to the fuel tank, where the low-pressure fuel pump takes suction. The engine has a fuel supply and return manifolds, with the return manifold having a small orifice to maintain constant pressure on the fuel injection system.

The injector in Figure 3.10 is cooled by the continual flow of fuel through it, except during the injection process. The injector is mechanically operated by a rocker arm pushing on the cam follower of the injector. The downward movement of the cam follower pushes the plunger down as well.

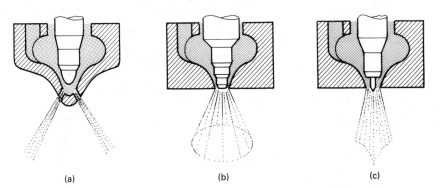

(a)   (b)   (c)

Figure 3.9 Nozzle spray patterns: (a) multihole; (b) conical pintle; (c) cylindrical pintle.

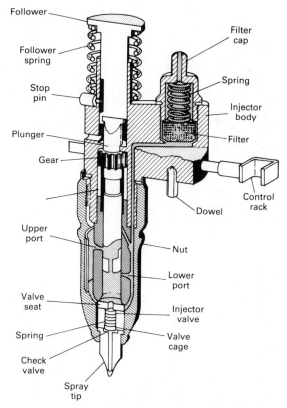

Figure 3.10   Unit injector.

The plunger, of the rotary type, is illustrated in Figure 3.11. There are two ports within the plunger barrel, the upper and lower ports. Within the plunger itself there is a central passage drilled from the bottom of the plunger up to the helix area.

Let us examine the sequence of operation as the plunger travels down the barrel. Motion begins when the rocker arm pushes on the follower, which is resisted by the follower spring, causing the plunger to move downward. At the top of the stroke, there is equal pressure throughout the plunger, as fuel flows in the upper and lower ports and through the central passages. As the plunger moves down, the lower port is covered, as shown in Figure 3.11a. When the helix covers the upper port, the pressure is no longer relieved from under the plunger and the start of injection begins. This is illustrated by Figure 3.11b. The plunger continues to move downward and the plunger barrel finally uncovers the lower port; this is the end of injection (Figure 3.11c). The plunger continues moving to the bottom of the stroke (Figure 3.11d). The plunger must travel the entire stroke (a) to (d) as it is positively displaced by the rocker arm. In this case the upper port controls the beginning of injection and the lower port controls the ending of injection. Often the unit injectors are variable beginning and constant ending. There are some, however, that have a slight helix on the lower portion of the plunger, causing a variable ending as well. The barrel and

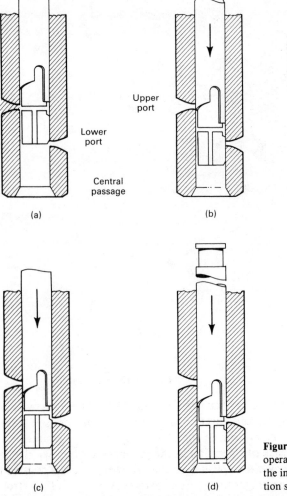

Lower
port

Upper
port

Central
passage

(a)   (b)   (c)   (d)

Figure 3.11 Sequence of unit injector operation: (a) top of the stroke; (b) start of the injection stroke; (c) end of the injection stroke; (d) bottom of the stroke.

plunger are not interchangeable, as they are a lapped set, with very close tolerances. If one is scored, the other will almost necessarily be scored. Even if it is not, the entire unit must be replaced. The governor moves the control rack, which rotates the plunger to give maximum fuel or no fuel, just as in the case of the fuel pump.

GOVERNOR

At the start of this chapter we noted that the fuel system had to deliver the fuel and control the engine's speed. The device that provides this control is the governor. The governor moves the fuel pump plunger, providing more or less fuel as the engine demands. There are several types of governors, the simplest being the mechanical one. There are others: hydraulic, pneumatic, and electronic. First, let's see what a governor should do. It depends on the application for which the engine is used.

There are constant-speed governors, typically used on diesel generators sets, where it is important to maintain a constant engine speed regardless of load. There are also variable-speed governors, which maintain a set output speed at any given load, although that speed may be varied with load through the use of the governor. Thus the governor will maintain a constant engine speed regardless of engine load changes. This type of governor does not meet with our expectations for vehicle use where we wish to have idling speed and maximum speed controlled but all other speeds set manually, by our foot on the "gas" pedal, the throttle. This type of governor is a speed-limiting governor. Its purpose is to prevent the engine from overspeeding when load is suddenly taken off the engine, and from stalling when the throttle is in the lowest possible position.

Another general function that governors provide is overspeed protection. These types of governors are used in conjunction with speed-regulating governors and there are two general types: manual and automatic reset. Figure 3.12a illustrates a speed–time diagram for the manual reset type, and Figure 3.12b shows the automatic reset type. The manual type will stop the engine if overspeed occurs, and the operator must reset the trip to start the engine. The automatic type does not stop the engine and the governor resets itself. The mechanism for stopping the engine is to eliminate the engine's fuel supply; the overspeed governor does not regulate fuel supply. Of course, an alarm would sound, notifying the operator of the problem.

Before proceeding further with governor types found on diesel engines, some terminology used in describing governor action needs to be defined.

Compensation: A mechanical and/or hydraulic action that prevents overcorrection of the fuel supply by the governor. Overcorrection causes the engine to fluctuate in speed.

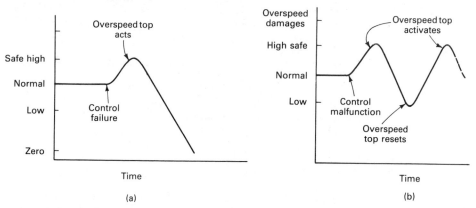

Figure 3.12   Speed–time diagram for (a) an overspeed trip governor and (b) an automatic reset/ overspeed trip governor.

Dead band: A narrow speed range, or band, in which the governor makes no correction in the fuel supplied to the engine.

High idle speed: The maximum speed at which the governor allows the engine to run.

Isochronous: Constant speed, regardless of load.

Low idle speed: The lowest speed at no load at which the governor controls the engine.

Overspeed: A speed above high idle speed.

Sensitivity: The smallest speed change that will cause the governor to change the fuel supply.

Speed droop: The difference between the high idle speed and the engine speed with full load, divided by the engine speed with full load. For mechanical governors this is 5 to 10%.

MECHANICAL GOVERNOR

Let's examine the simplest type of governor, the mechanical governor, illustrated in Figure 3.13. The mechanical governor is linked directly to the fuel pump's rack so that the action of the flyweights on the governor will cause the fuel pump rack to move, changing the amount of fuel that enters the engine. The governor is connected directly to the engine by a small driveshaft, which may be geared up to drive the flyweights at a speed greater than the engine speed. Such an arrangement is common on high- and medium-speed engines. The flyweights are attached to the ballhead, which meshes with the driveshaft gear. The flyweight's outward movement, due to centrifugal force, is resisted by a speed spring. The spring moves, and this change of

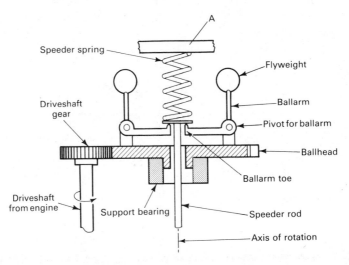

Figure 3.13 Mechanical governor.

position causes the speeder rod to move. The speeder rod is directly connected to the fuel pump. In the sketch shown, as the rod moves down, the fuel flow to the engine increases. In a simplified form, a connection from the throttle, or pedal, will increase the force acting on the speeder spring by pushing on plate A.

Let us analyze mechanical governors in more detail. Figure 3.14 illustrates a mechanical governor at various engine speed–load configurations. What must a mechanical governor consist of to operate? First, it must have a set of flyweights driven by the engine, typically at a higher rpm than the engine speed, to increase their sensitivity to small changes in engine speed; second, there must be a governor spring opposing the centrifugal force of the flyweights, allowing the governor to reach an equilibrium value; third, the flyweights must actuate, through a direct linkage, the fuel pump rack, which, in turn, meters the fuel to the injectors; and fourth, there must be input from the throttle control, changing the spring force. These components are shown schematically in Figure 3.14. In this case a pipe is used in lieu of a fuel

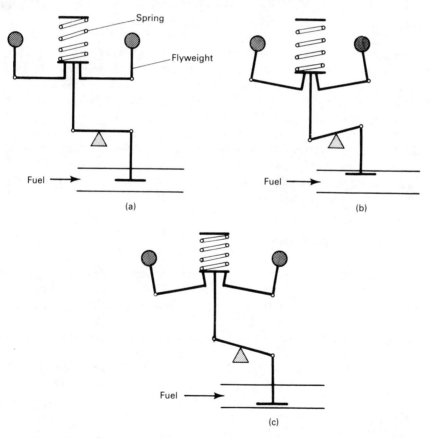

Figure 3.14 Flyweight variation with load for a mechanical governor: (a) equilibrium position; (b) increased load, increased fuel flow; (c) decreased load, decreased fuel flow.

rack, with the positioning of the fuel rack simulated by an opening or closing of a valve in the pipe, permitting more or less fuel low.

Figure 3.14a shows the equilibrium value of the mechanical governor at fixed load conditions. Assume that the engine is running at half load. Note that the flyweights are in the vertical position, with an equilibrium or static force balance existing between the governor spring and the flyweights. Assume now that the engine goes from half load to full load. The engine speed decreases. The flyweights move inward, as the centrifugal force of the flyweights at the lower speed does not balance the spring force. The result will be as shown in Figure 3.14b. The valve opens further, allowing an increased flow of fuel, so that the engine can run at the higher load. Note, however, that even when a new equilibrium is achieved, the flyweights will not be able to go back to the original vertical position which they had at half load. This is because the valve must be open further to allow greater fuel flow for the higher load. Thus the mechanical governor automatically has a speed droop. Now assume just the opposite—that the engine goes from half speed to low speed. The low speed is illustrated in Figure 3.14c. The flyweights move farther out as the throttle is shut and the engine increases in speed, shifting the valve to close off the fuel supply to the engine.

The flyweights must be large enough and/or the spring of the mechanical governor must be strong enough to cause the fuel pump rack to move. These requirements are adequate on small engines where the parts to be moved are also small and the force required by the spring or flyweights is not large. The inertia of all parts must be small to adapt to the inertia of the flyweights.

There is also a maximum fuel stop on the governor, which prevents additional fuel from being injected into the cylinder. Smoke limitations and engine maximum load conditions are two reasons for such a limitation. The engine can combust only so much fuel before smoking occurs; also, if too much fuel is burned, the heat release may be too great and damage the piston or cylinder walls.

HYDRAULIC GOVERNOR

Another type of governor used on today's diesel engines is the hydraulic governor. This type of governor reacts very quickly because it is hydraulic and the inertia of its parts is low. The mass and inertia of an engine are comparatively high; it does not react quickly to speed changes. In the design of the hydraulic governor, mechanisms must be built in to account for the differences in engine and governor reaction times so that they will act in accord with one another.

The total time lag from the time the hydraulic governor senses the speed change until the engine reaches a new or corrected speed may be separated into five steps.

1.   The delay in the governor in sensing the change in engine speed or load (manifested by governor flyweights changing velocity)

2.   A delay within the governor from the moment the flyweights sense a change until the governor reacts to the change

3.   The time from when the governor starts to change the fuel pump's setting (rotating plunger) until the fuel pump is charged

4.   The time it takes for the corrected fuel charge, once the plunger is reset, to enter the engine cylinder

5.   The time required for the correct fuel charge to be converted into the correct engine speed

Figure 3.15 is a simplified sketch of a hydraulic governor. It consists of three main parts: a speed-sensitive section, which senses the engine speed and tries to maintain constant engine speed; a power section, which actuates the fuel racks controlling the flow of fuel to the engine; and a compensating section, which compensates for the differences in reaction times between the engine and governor.

Let us consider Figure 3.15 in more detail and go through the times in the figure. The figure does not indicate two other parts of the governor body, a gear pump and accumulator. The gear pump is driven from the engine and supplies high-

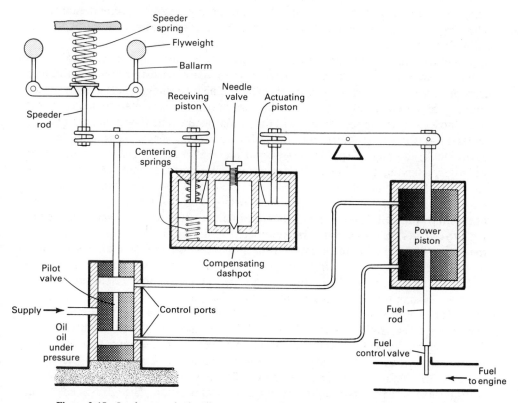

Figure 3.15   Isochronous hydraulic governor. (Adapted from Woodward Governor Co.)

pressure oil to an accumulator. The accumulator stores oil under pressure. Its pressure is constant, fixed by spring tension, which is determined by the manufacturer. The gear pump is a positive-displacement pump and will increase the oil pressure in the accumulator above its set point. The accumulator has an internal piston, resisted by the spring, that will rise, allowing oil to leak out through a port in the accumulator and back to the oil sump in the governor. The gear pump takes suction from this pump and the circuit is complete. The flyweights are also driven by the engine, typically at higher than engine speed. The power piston activates the fuel pump rack. It receives oil under pressure from control ports which are covered by the pilot valve when the system is in equilibrium. Should oil pressure increase on the bottom of the power piston, the piston will rise, allowing a greater fuel flow to the engine. In this case the control of the fuel pump rack is indicated by a valve allowing more or less flow through a pipe. The gear pump, accumulator, and power piston make up the power section of the hydraulic governor. The pilot valve, which is directly attached to the speeder rod, and the flyweights, make up the speed-sensitive section of the governor.

Somehow there must be a feedback mechanism within the governor so that it does not hunt. The compensating dashpot performs this function. Within the dashpot are located the receiving piston, the actuating piston, and the needle valve. The needle valve controls the flow of oil in and out of the closed hydraulic compensating system.

Let us consider the action of the governor for a constant-speed situation, such as when the diesel engine drives a generator. In this case the governor will maintain a constant speed for all loads. Assume that the engine is running at half load at a fixed rpm. In this case the only way an equilibrium position can be maintained is for the pilot valve to cover the control ports to the power piston. The flyweights must be in a vertical position for equilibrium to be achieved. Assume that there is an increase in engine load; the engine rpm decreases and the fuel rack linkage moves up in the governor, allowing more fuel to flow to the engine. Since the engine has decreased in speed, the flyweights will collapse, pushing the speeder rod down, which causes the pilot valve to lower, uncovering the lower port. The oil passage from the gear pump opens, allowing a flow of oil and an increase in oil pressure to the power piston, causing it to rise. As the power piston moves up, the actuating piston moves down, raising the pressure of the oil in the compensating system. This increase in oil pressure acts on the receiving piston, causing it to push the floating lever up, which in turn is attached to the pilot valve. This force pushes the pilot valve toward the equilibrium position, closing the port to the power section. Simultaneously, there is a leakage of oil from the compensating system through the needle valve back to the sump. As the engine speed reestablishes itself, the equilibrium position of the flyweights returns. The flyweights are vertical, and the pilot valve moves down, closing off the port from the power section. In this case the engine is running at a new and higher load; however, the engine speed remains constant. The port will be closed and equilibrium reestablished when the flyweights are in the vertical position. The adjustment of the needle valve is critical to the compensating system. The adjustment must be fixed such that leakage through the needle valve is coincident with the resumption of speed by the engine. This adjustment is initially made by the operator when the governor is installed and typically another adjustment is not necessary. The most common exception would be if dirt gets in the orifice of the needle valve.

Let us assume that the engine is operating at half load and that a change of load occurs; however, in this case the engine load is removed. Intuition tells us that the fuel rack leakage must move down, limiting the fuel supply to the engine. Let us observe what happens. As the load is removed, the engine will overspeed, causing the flyweights to move out. This will cause the pilot valve to move upward, allowing a discharge of oil to the power piston through the upper control port. Because the pressure on top of the power piston is greater than that below it, it pushes the fuel rack down, closing off the fuel supply to the engine. The actuating piston retards this motion, as the oil pressure in the compensating circuit is now below that of the surrounding sump. Oil flows in through the needle valve to the compensating circuit. The receiving piston tends to pull the floating lever and in turn the pilot valve down, covering the control port to reduce the flow of oil to the power piston. As previously, the only way for equilibrium to be established at the new no-load position is for the flyweights to be in a vertical position. The compensating circuit always acts to reduce the effect of the power section. This balances the differences in reaction times between the governor and the engine. The needle valve must be set for the peculiarities of each engine to balance these differences. This type of governor is isochronous, which can be a disadvantage at times, so a speed droop adjustment is placed on the governor. This is shown schematically in Figure 3.16.

Consider what happens when the speed droop adjustment arm is moved to the right, changing the location of the fulcrum. The speed droop adjustment changes the spring force that the centrifugal force must overcome to reach the vertical position. The flyweights must be in the vertical position for the pilot valve to be closed. At no load the speed droop lever will be tilted higher to the right and the speed droop cam slightly compresses the spring. Thus the centrifugal force on the flyweights must be greater. At full-load rpm with the lever in a higher position indicative of full fuel flow, the spring force is less, and less centrifugal force is needed to move the flyweights to the vertical position. Thus the force is not uniform over the speed range of the engine, and a variation of rpm from full load to no load will occur. If the speed droop cam is moved completely to the left, all speed droop is removed, as the cam does not change the speeder spring at all.

In comparing the hydraulic governor and the mechanical governor, note that the hydraulic governor is more complex; however, it may be used as an isochronous governor, and it allows the speed-sensing section to be very sensitive and independent of the power section. The hydraulic governor may be used as a variable-speed governor by the use of the throttle acting on the flyweight spring. The action of the governor would be the same as discussed, but the speed of the engine and flyweights would vary.

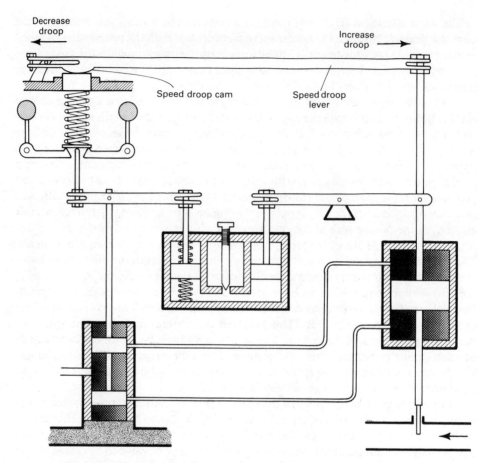

Figure 3.16 Isochronous hydraulic governor speed-droop lever. (Adapted from Woodward Governor Co.)

POLLUTION: PROBLEMS AND CONTROL

There are three types of exhaust emissions which have an adverse effect on the environment: hydrocarbons (HC), carbon monoxide (CO), and nitrogen oxides (NO_x). All three types occur in operating a diesel engine. The hydrocarbon emissions are due to not all the fuel being burned; some unburned fuel leaves the exhaust. This is different from soot or black smoke, which is carbon, a particulate matter formed by too much fuel for too little air. As part of the combustion process, the carbon in the fuel combines with oxygen to form carbon dioxide. In the process, however, some carbon and oxygen form carbon monoxide, which is then oxidized to form carbon dioxide. All this takes time, which in a high-speed engine is not plentiful, and some carbon monoxide is present in the exhaust. The last group is nitrogen oxides. Nitrogen is an inert gas and does not enter the combustion process, except at high temperatures. At high temperatures some of the nitrogen in the air will form nitrogen oxides.

Diesel engines are well designed to minimize the HC and CO emissions because air is controlled separately from fuel. This assures an ample supply of oxygen to readily combust the fuel, reduce HC, and oxidize CO. The nitrogen oxide formation is a more difficult problem, and is more a function of cylinder type.

The practice in reducing NO_x emissions in gasoline engines is to use exhaust gas recirculation. This is equally valid for diesel engines, as the idea to reduce peak combustion temperatures as well as limiting the oxygen available to form NO_x. Some of the exhaust gas is cooled and recirculated to the inlet air manifold. The engines have less oxygen, so the smoke limitation is nearer the actual fuel load conditions. This system works better with swirl combustion chambers than with the main chamber type of engine, as the increase in turbulence is required to keep any localized high combustion temperatures from occurring. Main chamber combustion engines suffer serious performance deterioration, with modest (10%) exhaust gas recirculation. This is not the case with swirl chamber engines.

It has been suggested to retard the fuel injection timing from an average of 21 degrees BTDC to 17 degrees BTDC. This was tested and while it caused a significant decrease in NO_x levels, the CO and HC levels rose.

The primary effort in diesel engines is to limit smoke from the engine during transient conditions, particularly acceleration. A diesel engine that emits smoke during steady operation needs a tune-up. However, during acceleration even a well-tuned engine may smoke if controls are not installed or timed to limit the fuel based on the air available. Consider this case. The throttle is moved ahead, or a load suddenly increases; the governor responds immediately by sending more fuel. But the engine speed has not increased to provide the necessary air. In the case of turbocharged engines, the turbine must receive hotter exhaust gases to compress more air so that all the fuel may be correctly burned.

The method that is used is to have a compensating circuit on the governor to prevent the complete rotation of the fuel pump plunger until the air supply has caught up with the fuel. This is done by measuring the air intake pressure versus the fuel supplied. In some systems part of the fuel bypasses the injector; the bypass closes as the air pressure increases. This is reasonably easily accomplished with turbocharged engines, as the fresh-air supply pressure increases with increasing load.

ELECTRONIC FUEL INJECTION

Efficiency and emissions of a diesel engine depend to a great extent on the timing of the injection. The optimum timing varies with engine speed, load, and operating conditions and constitutes a compromise between performance and acceptable emis-

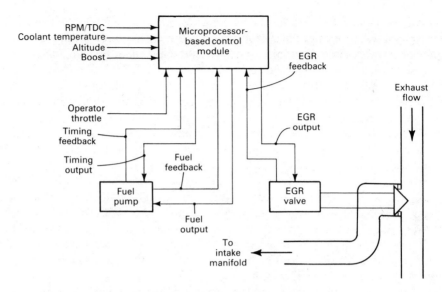

Figure 3.17 Diesel engine electronic control system.

sions. If, for instance, early timing is selected for best fuel consumption, unacceptably high NO_x emissions can result. In a general way, engine efficiency calls for earlier timing, whereas a reduction of NO_x emissions calls for later timing. To meet these conflicting requirements, timing has to be maintained within a very narrow band throughout all speeds, loads, and operating conditions of the engine.

The need is to achieve the best possible compromise between emissions, economy, and driveability. This requires very precise control of fuel injection quantity and timing. Electronics can achieve this far better than can hydromechanical systems.

For example, the change of timing under transient conditions can be much faster, permitting interlocking of the fueling and timing. Thus the timing is always correct for a given fueling, unlike present practice. This can be used to avoid the puff of smoke which so often occurs when accelerating from rest. Timing can also be modified as a function of a vehicle's acceleration rate.

With a fuel injection pump designed specifically for electronic control, the governing is better than with present-day pumps. The effect on idling speed of variations in load can be eliminated without the use of extra sensors or switches. Self-diagnosis of faults in sensors and actuators, and the electronics as part of a complete on-board diagnostics system, will also become increasingly important in the vehicle of the future.

Several types of electronic fuel injection systems are being developed at present. A schematic of the types of signals used is illustrated in Figure 3.17. Two signals return to the fuel pump: how much fuel to inject, the fuel output signal; and when to inject the fuel, the timing signal. The only other output signal is to the exhaust gas recirculating (EGR) valve, which allows the valve to open or close, varying the emission level in the engine exhaust so that it stays within prescribed limits. Such a system is quite complicated and is too complex for the present mechanical/hydraulic injection system. However, the present fuel injection pumps, which have proven so accurate and reliable, are still in use, but they are no longer cam-activated.

||||||||||||||||||||||||||||||||||||||||||||||||||||||||||||||||||||||||||||||||||||||||||||||||||||||||||||||||||||||||

Systems Necessary to Start and Keep the Engine Running

In this chapter we examine the engine support systems: cooling, lubrication, intake, exhaust, and starting. All of these must be kept in good condition for the engine to operate correctly. Let's start with the cooling system, which may be the most pesky, in terms of leaks and minor malfunctions.

COOLING SYSTEM

The function of the cooling system is to remove heat from the engine and maintain the engine parts at a nearly uniform temperature. We saw in Chapter 1 that about 26 to 30% of the energy of the fuel is removed as heat. This must be done at all load conditions, so the system must react to engine changes.

Most engines are water-cooled and we will examine this system in detail. There are air-cooled engines, as well, of several hundred horsepower. A major manufacturer is Deutz. In this system air is blown around a finned cylinder head, much like the fins on air-cooled motorcycles.

The water-cooled system is far more common. The water circulates through passages in the engine block and head, picking up heat from the engine. The hot water passes through an air-cooled radiator and is returned to the engine. There is more to it than this, but this in a nutshell is the system. Figure 4.1 illustrates a cooling system indicative of those found on larger diesels. The bypass line and the makeup line may not be present on smaller diesels.

The radiator seems like the most likely place to begin. When we think of the cooling system, it comes to mind right away. The radiator is a heat exchanger, but it is also a storage tank. When water is heated it increases in volume, the radiator must accommodate this expansion. The top of the radiator serves this purpose.

The radiator must be sealed from the outside and a pressure cap is used. It is not

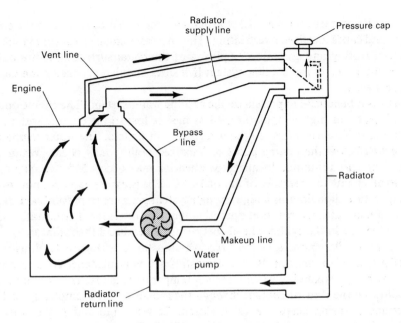

Figure 4.1 Cooling system.

desirable to have the cooling system have a vacuum on it, as this could cause air leaks into the system. In the hot engine environment the air bubbles cause erosion and pitting due to oxidation. Additionally, as the pressure on water is increased, its boiling temperature increases. Most pressure caps maintain a system pressure of 15 psi so that water boils at 250°F. Adding antifreeze raises the boiling point even higher. Figure 4.2 shows such a pressure cap. The cap has two valves, one the pressure valve, opening under high pressure, allowing vapor and sometimes liquid to flow out of the system via the overflow line. Additionally, there is a vacuum valve which opens to allow air into the system when the system cools down and may be under a vacuum. Sometimes an external expansion tank is connected to the overflow line, so a liquid seal is maintained on the cooling system at all times and air cannot normally enter.

Why is air so bad? It does not transfer heat well at all. In fact, it is a terrific in-

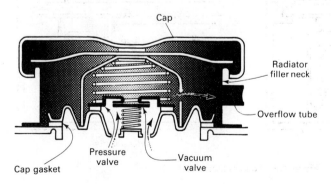

Figure 4.2 Radiator filler cap.

sulator. At times it can form a pocket around a component or section in the cooling passages and cause localized overheating. This increases engine wear dramatically. It also causes rusting and corrosion or pitting of the metal surfaces. Additionally, it displaces the water in the cooling system if it does circulate, reducing the engine's cooling capacity.

Why is it beneficial to pressurize the cooling water system? The engine operates more efficiently at high temperatures—less heat is lost to the cooling water as the temperature difference is less, so the heat flow will be less. The water temperature must be well below the boiling point or localized boiling may occur. When water boils, pitting may result and localized overheating may occur. Most cooling system maintain an engine temperature of about 185°F. The pressure cap does not regulate this temperature, it maintains a high enough system pressure to prevent boiling. The device that maintains the constant operating temperature is the thermostat, which is normally located in the cylinder head, between the head and the radiator.

Figure 4.3 illustrates a bellows-type thermostat, and Figure 4.4 illustrates a wax-pellet type of thermostat. Both thermostats act by a substance contained in the thermostat body, expanding when a certain temperature is reached, forcing a valve open, allowing the flow of coolant through the radiator. For example, up to 170°F the thermostat spring keeps the valve closed. Between 170 and 185°F, the valve should gradually open, being fully opened at 185°F. There are some thermostats, as shown in Figure 4.3, that always permit a small flow of coolant via the bypass line to the radiator.

If the coolant were always to flow through the radiator, the engine would not reach the desired operating temperature at low speed or light loads. This is detrimental in that the lubricating oil does not reach its temperature, which does not allow as

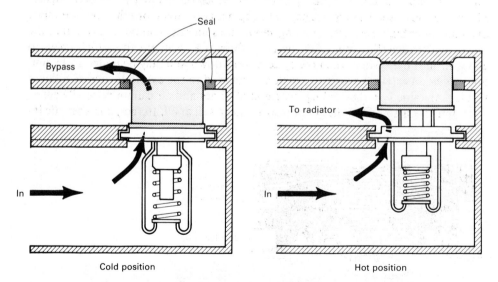

Cold position   Hot position

Figure 4.3 Bellows thermostat.

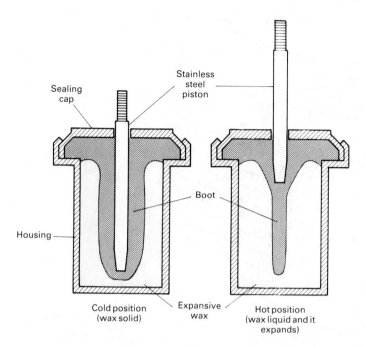

Figure 4.4 Wax-pellet thermostat.

effective an oil film to coat the engine's parts. Additionally, running a cool engine allows carbon to form on valves and injector tips.

In conjunction with a thermostat, some vehicles have shutters that allow or prevent air from flowing over the radiator. Figure 4.5 illustrates such a system. The *shutterstat* operates in a manner analogous to the thermostat, in that it senses the coolant temperature and transmits a signal to the shutter cylinder to open or close the shutters. The shutters minimize any airflow across the radiator when they are closed. This prevents the radiator from cooling and allows the engine to reach operating temperature in very cold weather. Obviously, the more moving parts a system has, the greater the likelihood of a problem. A decrease in airflow because of debris or stuck shutters becomes a problem in engine overheating in hot weather.

The water pump is the device that circulates the cooling water through the engine. It receives the coolant from the radiator, increases its pressure, forcing it through the engine, out through the thermostatic valve if it is open, and back to the top of the radiator. The pump is frequently either driven directly by the crankshaft, or driven by V-belts from the crankshaft pulley. It is a centrifugal type of pump, so the pump can rotate in the fluid without high-pressure buildup and without damaging the pump under no-flow conditions. The impeller is the last part shown in the water assembly in Figure 1.1. It is rigidly attached to the shaft and pulls water into the center, near the shaft, and flings it out from the impeller. When the water slows down, immediately after leaving the impeller, its pressure increases. However, the

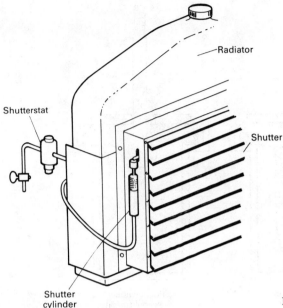

Figure 4.5 Radiator shutter system.

pressure can reach a maximum value only for a given speed; it will not increase indefinitely as would be the case if a piston were compressing the water.

Let's look at Figure 4.1 again and assume that the engine is running at the proper operating temperature. The coolant leaves the radiator via the radiator return line and goes to the water pump, which increases its pressure and forces it through the engine. The coolant flows through the thermostat and into the radiator via the radiator supply line. There is also a small fraction of coolant flowing to the pump through the makeup line. When the engine is cold, say just after starting, the thermostat is closed and water circulates through the bypass and back to the water pump. This allows the entire engine to warm up uniformly. There may be a flow of air and coolant to the radiator via the vent line.

Why is the makeup line there? To assist in filling the engine with coolant. The flow of coolant during filling is down the makeup line into the pump, the engine, and the radiator. As this happens, the displaced air flows through the vent line back to the radiator.

There is another item associated with the cooling system: the radiator fan or cooling fan. For stationary or slow-moving vehicles, something must be used to force air through the radiator grill. A fan is such a device. The fan may pull or push air through the radiator. It is thermostatically controlled, so it turns on only when the coolant temperature rises high enough to need it. There are several devices that perform this sensing operation. Should the fan be used in conjunction with a shutterstat, its set point should be several degrees higher than the shutterstat so that it will not start until the shutters are open.

The condition of the coolant is as important as there being an adequate amount of coolant. Very often a coolant filter is used to remove solid contaminants. These can be introduced during repair, filling with coolant, or rebuilding. The acidity or alkalinity of the water needs to be controlled. The acids in water, in particular, attack metal surfaces. Very often the engine will need antifreeze, which should be compatible with the filter. The antifreeze not only lowers the freezing point, but raises the boiling point of water. Additionally, antifreezes have inhibitors in them which coat coolant surfaces, preventing oxidation (corrosion) and forming a protective shield for erosion. Erosion is caused by cavitation at the liner or engine surface. This occurs when liner vibration causes vapor bubbles to form; cavitation, in effect, wears away the metal lining. When the inhibitor has coated a metal surface, the cavitation effects act on the inhibitor rather than on the metal surface. The inhibitor is maintained in the water, so as it is depleted by cavitation, it is replaced by chemical coating. Additionally, the coolant additives prevent the formation of scale, which reduces heat transfer and causes hot spots on the interior metal surface.

LUBRICATION SYSTEM

As soon as we begin to start an engine, wear begins. It is vital that moving adjacent parts be separated by an oil film. Even the smoothest metal surface consists of many surface irregularities (peaks and valleys) when examined under a microscope. The oil film fills in these irregularities and prevents the two metal surfaces from touching one another. On startup, particularly if the engine has not been run for a while, the oil film may not be in place and momentary metal-to-metal contact will occur, shortening the engine's life. Additionally, if the engine is operated at overload conditions for a sustained period, the higher temperature of the engine parts may result in the metal expanding beyond its normal limits. Metal always expands when heated, but a certain space between metal parts, a *tolerance,* is designed for. Should the metal expand beyond the design limits, the metal surface irregularities may touch as the oil is squeezed from between the surfaces.

Figure 4.6 illustrates an oil system found on truck engines. A lubricating oil pump is driven from the engine and provides more than sufficient oil at a pressure high enough to reach the most remote moving part. Let's follow the system through its various components. The oil pump is a gear-type pump. The oil is drawn into the pump through a screen or strainer, located near or in the lowest point in the oil pan or sump. Figure 4.7 illustrates a gear-type pump. Oil enters the pump, is carried around the periphery in the gear teeth, and is discharged. The oil is prevented from returning to the inlet side by the meshing of the gears. The oil pressure will increase to whatever amount is necessary by the meshing gear teeth. The oil flow rate is dictated by the engine speed. However, even at low speeds the pump must deliver an adequate supply of oil. Also, as the engine wears and clearances are greater, the oil supply must be adequate to meet this condition.

An oil cooler follows. The cooler bypass assures a flow of oil to the system if the

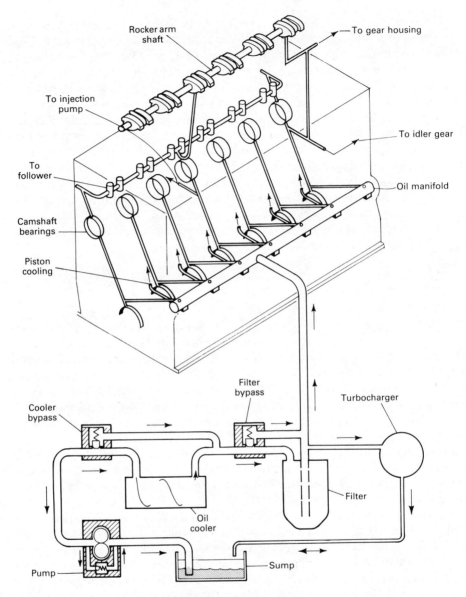

Figure 4.6 Lubricating system.

cooler becomes blocked. The oil cooler uses water from the radiator as its cooling medium under operating conditions. Figure 4.8 illustrates such a cooler. Not all the engine heat is removed directly by the coolant; a significant portion is removed by the lubricating oil circulating through the engine parts. The hot oil returns to the oil pan and some heat is removed by the airflow around the oil pan, but not all. The oil's

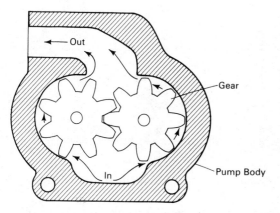

Figure 4.7 Gear-type oil pump.

lubricating properties are temperature dependent and break down under high temperature. We see that the hot oil flows through the tubes, with the coolant from the radiator flowing around them. Heat always flows from hot to cold, and in most instances, but not all, the oil is cooled. During engine startup the cooling water will increase in temperature faster than the lubricating oil does, and during this time period the cooling water will increase the oil's temperature. Remember that the thermostat did not allow the cooling water to flow through the radiator until reaching its operating temperature. This further shows the advantage of using a thermostat. In terms of temperature, most coolant temperatures range from 160 to 185°F, while oil-return temperatures to the sump are up to 250°F.

Lube oil coolers are the place where contamination of the lube oil or cooling water most frequently occurs. Both are detrimental to engine life, as water and lube oil form a sludge. While on the topic of problems, when the engine, particularly if it is turbocharged, has been running under rated load or speed, it should be allowed to idle for some time before being stopped. This is because the metal surfaces become

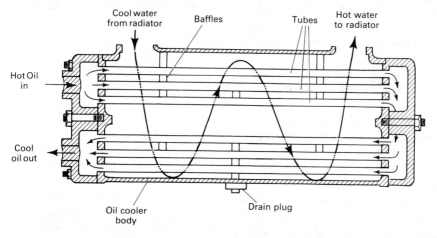

Figure 4.8 Oil cooler with water and oil paths.

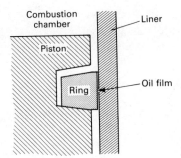

Figure 4.9 Ring-oil contact on the liner surface.

very hot under load and shutting the engine down quickly means that oil stops circulating and burns on hot surfaces.

The next component is the filter, which often uses a pleated paper as the filtering agent, although other media are also used. The engine always needs a flow of oil. Should the filter become clogged, allowing a marginal flow when the demand requires greater flow, the bypass valve is used. It is shown separately in this sketch, but it may be a ball check valve on the top of the filter housing. In any event, if the pressure drop across the filter is too great, the bypass valve opens. This is not good for the engine, as dirt and contaminants will flow to the engine surfaces.

The oil then proceeds to the engine via the oil manifold and to the turbocharger. The return line from the engine (not shown) brings the oil back to the sump. This is often the crankcase. The engine shown here provides for piston cooling, which is important at high loads when the piston comes under high heat loads.

The lubricating system has four main purposes: to reduce friction and wear in the engine, to clean the engine interior of dirt and carbon, to cool the internal parts of the engine, and to act as a sealant between the rings and liners.

The first two purposes lead us to an understanding of oil properties, which will soon follow. The third purpose we have seen by examining the lube oil system, and the fourth is illustrated in Figure 4.9. The oil seal formed by the contact of the piston ring with the liner prevents combustion gas on the power stroke and intake air on the compression stroke from entering the crankcase.

OIL: PROBLEMS AND PROPERTIES

The lubricating oil is required to reduce friction, to clean dirt and carbon from the engine, and to maintain the correct viscosity (to maintain its antifriction characteristics) at a variety of temperatures. Contamination of the lube oil is the severest problem. The oil does not wear out, but it becomes contaminated to the point where it cannot function correctly.

What are these contaminants, and where do they come from? They are dirt, water, dust, other abrasives, suspended metal particles, diesel fuel, gases, and vapors. Water can decrease the oil's viscosity, which can endanger the metal surfaces when they touch. In large amounts in the sump, it will give a false oil level and will react with oil to form a sludge, which can clog the drilled oil passages and leave deposits in the system. Additionally, it can contribute to the rusting of metal parts during shutdown, because they are not protected by the oil film. A typical place water enters the system is through condensation when the outside temperature is low. When the temperature is below the dew point, water will precipitate from the air in the crankcase. Also, a leak in the lube oil cooler is a likely place for water infiltration. Dirt, dust, and abrasives can enter the system during oil addition or engine repair. Their negative effect in increasing wear because of their scouring nature is apparent. Metal particles, particularly small ones, occur naturally, especially during engine break-in. There is some wear occurring all the time, and a ferrographic analysis of the oil can indicate what part type is wearing and at what rate. This is used as a diagnostic tool in engine maintenance.

Diesel fuel will enter the lube oil if it leaks around injector O-rings during incomplete combustion. Fuel dilution may occur with excessive stop-and-go operation. This is a less damaging contaminant than others, but it does not provide the same viscosity protection as lubricating oil.

Figure 4.10 illustrates blow-by of a piston ring. Such blow-by could also exist between valve and valve guides and turbocharger seals. The gases contain particles of carbon, water vapor, acids, partially burned fuels, varnish, and lacquers. Also, the oil itself may oxidize when coming in contact with the hot metal surface. Remember the oil scraper rings on the piston. They remove excess oil from the liner back to the crankcase. This oil will have some contaminants in it under the best of circumstances.

How does the oil handle such problems? Well, the oil is more than oil, but rather, an oil base stock to which chemicals are added to overcome or minimize the contamination problems. It is this combination of additives plus high-grade base oils that meet the engine requirements. What types of additives are provided:

Viscosity improvers: Generally, temperature-sensitive polymers. When the oil is cold they coil tightly and have little effect on viscosity. As the oil increases in temperature, the polymers uncoil and interact with one another, thickening the oil.

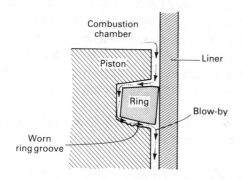

Figure 4.10 Illustration of piston ring blow-by.

Antifoam agents: Generally, silicone polymers. A large amount of air in oil causes foaming. These polymers attach themselves to small air bubbles, and combine, forming a large bubble rising to the surface. Since fewer bubbles reach the surface, there is less foaming.

Friction modifiers: Chemicals that provide an additional protective film on metal surfaces, as the oil cannot, by itself, maintain a sufficient oil film under extreme pressure.

Detergents and dispersants: Additives that attach themselves to carbon, partially burned fuel, lacquers, and varnish, and hold them in suspension, preventing sludge deposits. These particles are so fine that they cannot be removed by filtration. The oil must be changed. Should the contaminant level become too great and all the detergent/dispersant be used, the additives "overload" themselves and fall out of solution, forming a sludge with the particles to which they attached themselves.

Oxidation inhibitors: Additives that stop the oxidation reaction of lube oil. When the lube oil comes in contact with a hot engine spot, oxidation occurs, creating oil contaminants. These contaminants promote more oxidation; the inhibitor acts to prevent this.

Corrosion inhibitors: Generally, alkaline materials that neutralize the acids formed by the products of combustion, carbon and sulfur oxides, combining with water. The acids rapidly corrode bearing and journal surfaces.

Antirust inhibitors: Inhibitors that form an insulating film on materials, preventing contact with water.

The oil must be replaced before the additives are depleted. This varies with engine type and use. Manufacturers given recommended maintenance scheduling for different types of service. Additionally, oils are graded according to their use.

The American Petroleum Institute, in cooperation with several other engineering societies and manufacturers, has developed a letter designation for various engine operating conditions and the type of lube oil necessary.

SA: A lubricating oil used for gasoline and diesel engines when the engines are operated under such mild conditions that a compounded oil is not required.

SB: A lubricating oil designed for medium-duty gasoline engine service. It provides antiscuff capability and resistance to oil oxidation and bearing corrosion.

SC: A lubricating oil designed for gasoline engines. It provides control of high- and low-temperature deposits and gives protection against wear, rust, and corrosion.

SD: A lubricating oil designed for gasoline engines. It gives greater protection against high- and low-temperature engine deposits, wear, rust, and corrosion than an SC oil.

SE: A lubricating oil designed for modern gasoline engines. It provides greater protection against oil oxidation deposits, low- and high-temperature deposits, rust, and corrosion than do SD or SC classed oils.

CA: A lubricating oil designed for light duty, normally aspirated diesel engines. It protects against bearing corrosion and high-temperature deposits.

CB: A lubricating oil designed for moderate-duty diesel engines. It provides the necessary protection from bearing corrosion and from high-temperature deposits in normally aspirated diesel engines using fuels with higher sulfur content.

CC: A lubricating oil designed for moderate- to severe-duty diesel and gasoline engines. The engines are often ones used in trucks, industrial and construction equipment, and farm tractors. It provides protection from high-temperature deposits in lightly supercharged diesel engines and also from rust, corrosion, and low-temperature deposits in gasoline engines.

CD: A lubricating oil designed for severe-duty diesel engines. It provides protection against bearing corrosion and from temperature deposits in supercharged engines when fuels with a wide range of quality are used.

How are these classifications used? An engine manufacturer will specify a certain type of oil, depending on the service of the engine. For instance, Cummins Engine Company recommends an SC lubricant for their normally aspirated engines in stop-and-go service. The same engine when used in heavy-duty service requires a CC lubricant. Thus the type of oil used will afford the correct engine protection only when it is matched to the engine operating conditions.

There is another designation that is used in defining lubricating oils, that of *viscosity index.* The viscosity index tells the rate of change in an oil's viscosity as its temperature changes. There are high-viscosity index (HVI) and low-viscosity index (LVI) oils. The higher the viscosity index, the less the oil's viscosity changes between the temperature limits of 104 and 212°F. The MVI lubricating oil has been used in medium-speed engines, such as railroad engines, for quite awhile. There are differences in the kinds of molecules that make up HVI and MVI oils. HVI oils come from paraffinic lube crudes. They consist mainly of hydrocarbons that tend to produce hard carbon deposits when burned and decomposed in the combustion chamber area.

On the other hand, MVI lubes come from naphthenic crudes. MVI oil hydrocarbons form soft, crumbly carbon deposits when they burn and decompose from heat. This difference in molecular composition gives MVI oils the performance edge over HVI oils, particularly in two-stroke diesel engines.

Another factor is that HVI oils have a residual oil base stock, whereas MVI oils do not require this. The result is that the HVI base stock forms more and harder carbon deposits than those formed by MVI oils. In two-stroke diesel engines the formation of carbon deposits in piston grooves and exhaust ports is very detrimental, and MVI oils hold the advantage over HVI oils in this regard.

The relatively soft piston groove deposits from the combustion of MVI oils tend to be removed by the action of the rings. HVI oils tend to leave more dense and adhesive deposits that build up in the piston ring grooves. In time these harder deposits can prevent the ring from compressing completely into the groove, causing the ring to protrude from the piston surface when the piston is in the cylinder. Of course, scuffing of the cylinder wall can occur and broken or chipped rings also result.

Port blocking is particularly critical in two-stroke engines with exhaust ports, because it can affect engine power by increasing the pumping loss on the exhaust stroke. Fortunately, deposits formed in the exhaust ports from MVI oils slough off regularly, retarding excessive carbon buildup. HVI oils do not slough off, and keep building in the exhaust port area. This requires hand scraping to clear the exhaust ports between normal overhaul times. The result is increased downtime and maintenance cost.

The MVI oils do not have the high-temperature oxidation stability that HVI oils have, although this limitation occurs at 250°F. They are used on medium-speed engines which have a regulated cooling supply to the engine, assuring no high-temperature excursions in the engine.

INTAKE AND EXHAUST SYSTEMS

The air that is supplied to the engine must be clean, as any dirt or debris will reduce engine life and in extreme conditions cause piston seizure. The air intake system cleans, compresses in many instances, cools, and silences the air. Very often diesel engines are turbocharged or supercharged, resulting in compressed air flowing to the cylinders. Figure 4.11 illustrates an air intake system.

In larger diesels the pulsating airflow may cause vibrations and noise. An intake silencer that breaks up the sound waves may also be incorporated before the air cleaner. On most small high-speed engines, the air cleaner also serves as a silencer. Figure 4.11 illustrates a dry-type cleaner, but let us first examine an oil bath cleaner, which has been used for many years on engines, but is now yielding to dry-type filters.

Figure 4.12 illustrates an oil bath cleaner. The air, containing dirt particles, enters from the top and proceeds down, changing direction when it reaches the bottom and contacts the oil surface. Larger particles cannot change direction quickly and fall into the oil and settle to the bottom. The smaller particles pick up minute amounts of oil on their surfaces. The air and oiled dirt particles now pass through a wire mesh, which has been lightly oiled. The oily dirt particles adhere to the mesh and the cleaned air passes to the engine. The oil with most of the entrained dirt particles drips back to the bottom.

There are four potential problems with using this type of filter. The first is that of a dusty engine operating environment. The fine dust will pass through the wire mesh, and in some instances there is just too much dirt for the oil to remove. A dry-type filter should be used. The second and third problems have to do with pumping of the oil and dirt through the wire mesh, eventually rendering the cleaner useless. Water may collect in the bottom of the oil bath, raising its level. The water may be en-

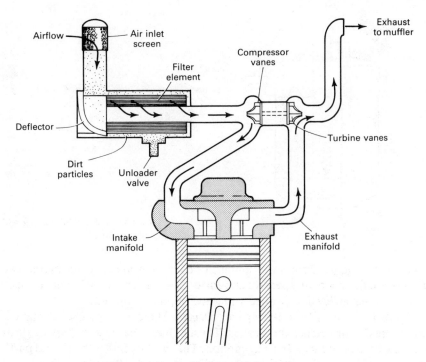

Figure 4.11   Turbocharged air intake system.

trained in the air, or from extremely humid air. Figure 4.13 illustrates this situation. The oil level is raised and the air may push it through the mesh. Also, a very high airflow rate can create the same effect, forcing the oil and dirt through the mesh and into the engine. The fourth problem lies with too little airflow, which can occur at idle speeds. The flow is slow, so it does not agitate the oil surface and the dirt particles will not have an oil film adhering to them. Thus the fine dirt and dust particles can pass

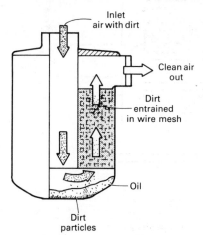

Figure 4.12   Oil bath air cleaner: normal operation.

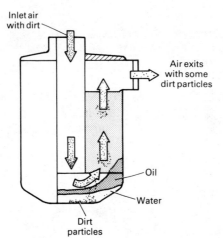

Figure 4.13 Oil bath air cleaner with oil and dirt carryover.

through to the engine. Prolonged idling in dusty environments can hasten engine wear. Oil bath filters are inexpensive and easily cleaned. The mesh is cleaned in a solvent, or steam-cleaned, lightly oiled with lube oil, and returned.

Figure 4.14 illustrates a dry-type air cleaner. There is, of course, a course inlet screen before the air reaches the cleaner. The air enters the cleaner, passing through fins or vans which produce a swirling motion. The swirling forces the air and particles to change direction, which the larger particles cannot do. They drop from the air-stream and collect at the bottom of the inlet cleaner section. In this illustration an exhaust educator (aspirator) pulls the debris into the exhaust system, where it is discharged into the surroundings. The educator works on the principle of creating a

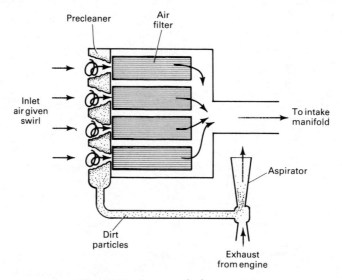

Figure 4.14 Dry-type air cleaner system.

velocity increase in the exhaust gases; as the velocity increases, the pressure at that location decreases to the point where the inlet air pressure is greater, and this pressure difference forces the dirt into the exhaust stream. The exhaust is slowed down in the wide part of the educator and the pressure returns to its original value.

The swirling air enters the filtering element, usually pleated paper held firmly in a metal frame. The filter can remove dust particles of very small micron diameters. By measuring the pressure difference across the filter, the filter may be replaced when the difference exceeds its recommended value. The pressure difference indicates that the filter is clogging with dirt particles and restricting the flow of air to the engine. This can cause smoking of the engine, as insufficient air is being supplied. As dirt builds up on the filtering element, the filter removes finer particles than necessary, creating a pressure drop in doing this. As the air's pressure decreases its density decreases, so less air mass enters the cylinder. The dry cleaners are very efficient at all speeds; however, for engines operating in construction, mining, or irrigation areas the air cleaner should be larger than normal to minimize the pressure drop. Also, the exhaust discharge should not be located near the intake, to minimize dirt and carbon in the exhaust gases from entering the intake system.

It is possible to use the energy of the exhaust gases to drive a turbine and have this turbine drive a compressor, which is attached to it by a common shaft. Figure 4.15 illustrates this. As soon as the engine begins to fire, the exhaust gases flow through the turbine volute, moving the turbine wheel. The compressor wheel turns, forcing more air into the engine. The greater the engine load and speed, the greater will be the turbocharger rotation. The compressor increases the pressure and temperature of the inlet air. Additionally, the temperature is raised by heat transfer from the hot turbocharger housing, which was heated by the high-temperature ex-

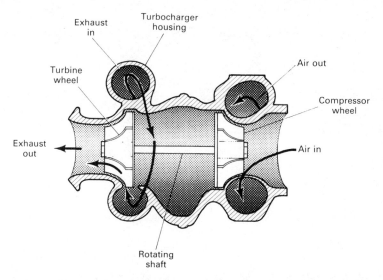

Figure 4.15   Turbocharger assembly with exhaust and airflow.

haust gases. This high-temperature air needs to be cooled, hence an *aftercooler* or *intercooler,* as it is sometimes called, is the next component in the air intake system.

Before we look at the aftercooler, why are engines turbocharged in the first place? A naturally aspirated engine runs perfectly well without turbocharging. Turbocharging allows a greater mass of air to enter the cylinder; thus, more fuel can be burned, providing more energy released in the cylinder, increasing the power potential of the engine; sometimes 50–70% greater power.

Turbocharging an engine causes a higher average pressure to act on the bearings. These are not necessarily higher peak pressures, but a higher sustained pressure on the bearing. This increase in pressure requires that the oil flow be increased compared to a naturally aspirated engine, to maintain the proper bearing temperature. The heat release in the engine is greater, so the cooling load is greater. One way this total cooling load is reduced vis-a-vis a naturally aspirated engine is by extending the overlap period for the exhaust and intake valves. Figure 4.16 illustrates a polar timing diagram for a self-aspirated engine with dotted lines superimposing the location of the intake valve openings and exhaust valve closings for a turbocharged engine. This extended overlap period allows for cooling of the exhaust valves, piston crown, and cylinder walls by the air system; hence, some of the cooling responsibility that might fall on the cooling system is taken care of cooling effect of the air.

Figure 4.17 illustrates an aftercooler arrangement using cooling water to reduce the air temperature and increase the density, hence, mass of air entering the cylinder. The air passes over the cooling water which circulates through the tubes. It is the responsibility of the cooling system, discussed earlier, to remove this heat.

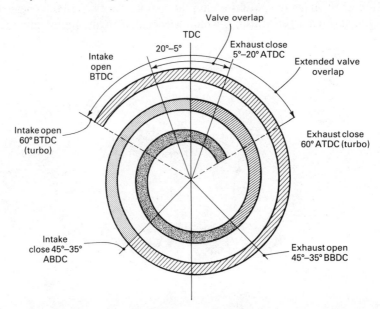

Figure 4.16   Differences in polar timing diagram between naturally aspirated and turbocharged engines.

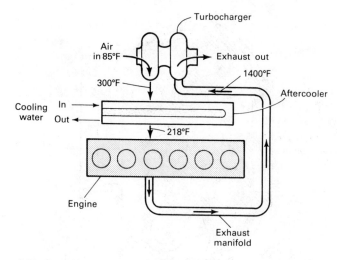

Figure 4.17 Temperatures in a turbocharged diesel engine intake and exhaust system.

It is not necessary that the aftercooler be water cooled. Figure 4.18 illustrates one that is air cooled. Some of the air from the compressor drives a fan which blows ambient air over a finned air-to-air heat exchanger.

Two-stroke cycle engines must always have a supply of air under pressure for the scavenging process. A supercharger is often used, Figure 4.19. There are driving

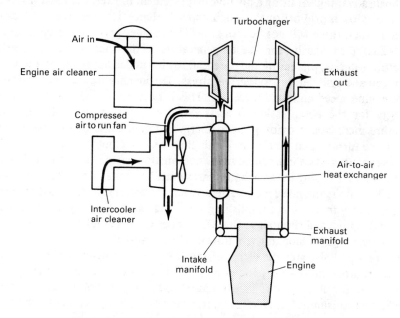

Figure 4.18 Air intake system with air-to-air intercooling.

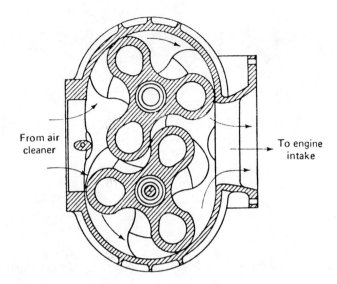

Figure 4.19 Roots-type blower.

gears on one end of the blower, driven from the engine. The opposite end has a lobe type compressor (the end shown in the sketch). The lobes force the air through the pump; it is forced out of the discharge at a higher pressure. The gears maintain the spacing between the lobes so there is no metal-to-metal contact.

The disadvantage of using a positive displacement blower, such as a Roots type, is that the air flow is proportional to engine speed. Hence, if the engine overloads and burns more fuel, there will not be more air. The engine speed may even decrease at heavy loads, particularly for diesel generators. This means that the exhaust temperature will rise, increasing the potential for burned exhaust valves and damage to the piston and cylinder due to thermal stress. Turbocharging increases the air supply with engine speed and load increases. This is because the exhaust gases provide the energy for the compression process. As the engine is loaded, the exhaust temperature increases, providing more energy per unit mass of exhaust to be extracted in the turbine and, hence, more air that may be compressed.

The exhaust system must remove the combustion gases in a controlled manner. The noise waves traveling in the exhaust gases must be broken up before reaching the open air. This limits the noise we hear. However, the exhaust back pressure must not be increased very much in accomplishing this. The exhaust piping and components have to be designed and situated so that water does not accumulate, causing rust and in extreme cases, gases blocking the flow of the exhaust gases.

Figure 4.20 illustrates the effect of increased back pressure on engine performance. More work must be expended on pushing the exhaust out of the engine, so less is available for useful work. This is manifested as a loss of power and an increase in specific fuel consumption. Additional indications are a high coolant temperature,

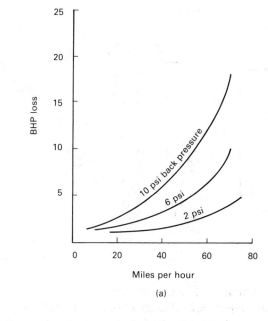

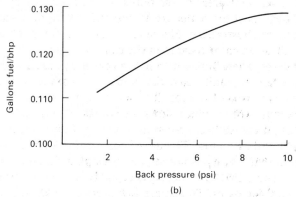

Figure 4.20   (a) Horsepower loss versus back pressure. (b) Increase in fuel consumption with increase in back pressure.

because of increased cooling required due to higher pressure and temperature exhaust gases. Also, engine wear will increase due to the higher average temperatures.

Mufflers are used to reduce the noise of the exhaust gases by the use of internal baffling. The velocity of the gases decreases, and the pressure increases. The more effective the silencing, the greater the pressure rise. So muffler selection by the engine manufacturer is a compromise between noise reduction and back-pressure increase.

The tailpipe is that portion of the exhaust system after the muffler. Usually the exhaust system terminates with a bend or a cap at the end of the tailpipe. Drain holes

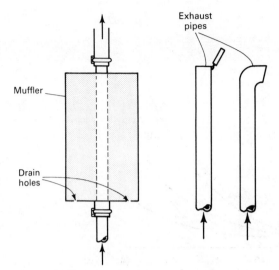

Figure 4.21   Exhaust pipes.

on the muffler may eliminate water that has come into the system, but it is better not to have it enter at all. Figure 4.21 illustrates this. Caps are normally recommended because bends under 90 degrees at the outlet will not stop water entry and bends greater than 90 degrees at the outlet are against the law in some states.

Exhaust brakes, illustrated in Figure 4.22, may be installed on some vehicles. A butterfly valve arrangement is located in the exhaust piping, which can restrict the flow of exhaust gases. These brakes help slow a vehicle by reducing the engine power. Since the exhaust cannot readily leave the cylinder, less oxygen is available for combustion on the power stroke. Additionally, there is greater pressure acting against the piston on the exhaust stroke, which slows the crankshaft rpm and acts as a brake.

An addition to the exhaust system in future light duty diesel engines will be a device to trap particulate matter. The use of combustion chamber design modifications is one avenue that is being explored, but the emphasis is on improved efficiency. Complicating matters is that by achieving acceptable NO_x levels, usually with EGRm the particulate level increases. To achieve low particulate levels in the exhaust, 0.2 grams per mile, a regenerative type filter will be required.

Corning Glass has developed a ceramic filter to meet the filtering requirements

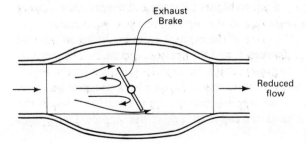

Figure 4.22   Exhaust brake.

of a diesel engine exhaust system. A honeycombed ceramic filter is used which forces the exhaust through the ceramic wall. The particulate matter is left on the wall. The problem of increased back pressure is obvious. Since an automotive diesel engine will emit a gallon of particulate matter for every thousand miles driven, there must be some way that the particulage can be removed without disassembling, or replacing the filter. This is the regenerative feature. If the exhaust temperature in the filter is raised to 850°F the particulate that is soot will oxidize. The ceramic filter holds the potential to withstand the high thermal stress caused during this regenerative period. There is experimentation with separately fired filters or with temporary engine operating conditions which will cause the temperature to reach 850°F.

STARTING SYSTEMS

Thus far the diesel engine has always been running, but it is necessary to start it. The engine must be turned over with sufficient speed so the temperature of the air in the cylinder is high enough to ignite the fuel when it is injected. Pressure and temperature go hand-in-hand, and Figure 4.23 illustrates the relationship between rmp and compression pressure. A pressure of about 400 psi is required to produce a high enough temperature to ignite the fuel. At lower cranking speeds there is too much time for heat loss for the cylinder to build up the necessary temperature and pressure. When the outside air temperature is low it is difficult to attain the required temperature without preheating of the air by some means. Further explanation follows later.

There are several ways in which diesels may be started. For automotive and truck use an electric starter motor is most common. However, other systems are commonly used on larger high speed and medium speed diesels. For instance, hydraulic motors, pneumatic motors, and small gasoline engines are all used to bring the diesel up to firing speed.

Let's look at the battery operated electric starting motor first. Figure 4.24 il-

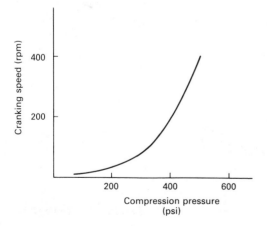

Figure 4.23 Compression pressure versus cranking speed.

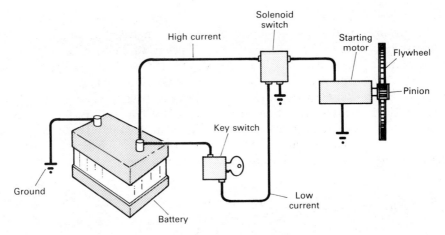

Figure 4.24 Simplified starting circuit.

lustrates the system in schematic form. The key switch allows current to flow to the solenoid switch, which in turn, allows the flow of electricity to go to the starting motor. The motor rotates and a small gear (*pinion*) engages with the flywheel, thus rotating the engine. Figure 4.25 illustrates a starting motor. The *solenoid* is an electromagnet; the flow of electricity through the solenoid causes it to exert a magnetic pull on the plunger within its core. The closing of the solenoid switch allows a flow of current to the starter motor. As it turns it rotates the pinion which moves down the motor shaft and engages with the flywheel, rotating the engine. When the engine starts, it causes the flywheel to drive the pinion faster than the motor rpm. The pinion moves out of mesh and current through the starter motor is stopped by the person using the key. The grinding noise sometimes heard when a person tries to start an engine already running is that of the pinion hitting the flywheel and not engaging.

A problem using batteries as the energy source is one, that they deplete with use and must be replaced; and two, their power falls off rapidly with decreasing temperature. Figure 4.26 illustrates this. However, the battery/motor starting system is very compact and easy to use.

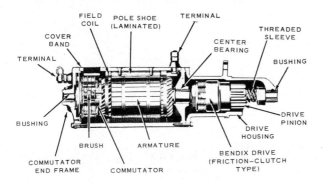

Figure 4.25 Starting motor.

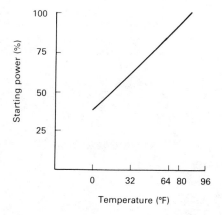

Figure 4.26 Effect of temperature on the battery starting power.

The hydraulic starting system uses a hydraulic motor rather than an electric motor. Figure 4.27 illustrates this type of system. The accumulator holds a quantity of oil under high pressure, 1500–3000 psi. When the engine is to be started, the oil flows from the accumlator to the hydraulic motor, causing it to rotate. The low pressure oil goes to the reservoir. The starter motor has a pinion which engages with the flywheel. When the starting lever is released the pinion is disengaged and the flow of oil stops. Since oil is not very compressible, a nitrogen charge is placed at one end of the accumulator, it expands as the oil leaves the accumulator maintaining a decreasing, but high pressure.

The oil must be returned to the accumulator once the engine is running and the engine driven pump does just this. In addition a hand pump is included in the system

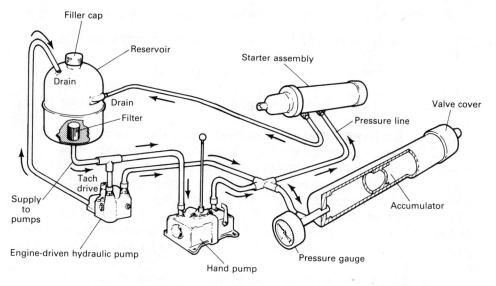

Figure 4.27 Hydraulic starting system.

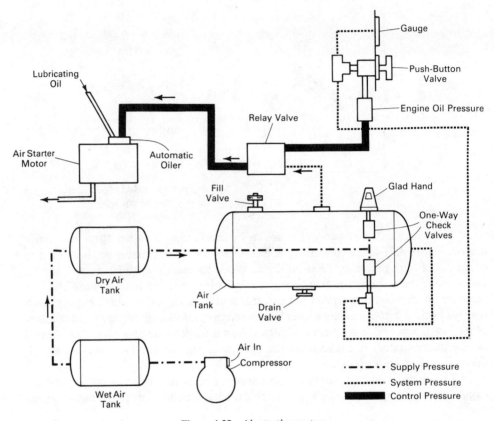

Figure 4.28 Air starting system.

as a back-up, or if the pressure in the accumulator is not high enough for engine starting.

A pneumatic motor system is illustrated in Figure 4.28. The air compressor supplies both the air brake system and the starting air system. The starting air system must have its own tank. A push button valve activates the flow of air the air starting motor, via the relay valve. When the operator releases the push button valve, the relay valve is inactivated and air no longer flows to the starter. The air starting motor operates in an analogous manner to the hydraulic motor. In this case, however, the air is not recycled but vented to atmosphere via a small muffler to reduce the noise. There is no hand pump for emergency starting or supplying compressed air if the engine compressor is stopped. In this case a glad hand is provided. It is a fast break away coupling which allows hook-up with another system via an emergency air supply line. The air does not have lubricating qualities which all rotating machinery needs—so an external source of lubrication is provided for the motor—the automatic oiler.

There are several advantages for the air and/or hydraulic starting systems vis-a-vis the electric systems. The cost and maintenance are less for the air system, about one-third that of hydraulic and electric systems. Both systems can crank the engine faster than the electric system. The electric system may crank longer, but we have seen that speed is what counts in building up the air temperature. The air system, rather than the hydraulic system, has the potential for longer cranking time, but adding more accumulators balances this point. The air and hydraulic systems are not temperature dependent in their power supplied to the starting motor, while the battery certainly is.

The ease of engine starting depends on the surrounding air temperature. We have seen that oil viscosity increases with decreasing temperature, making the engine harder to rotate as the temperature drops. So it requires more starting power just when the battery can deliver less. Also, the metal parts are cold at low outside temperatures, necessitating a longer cranking period to bring the cylinder air temperature up the fuel's ignition point. What to do?

Frequently, *glow plugs* are used to preheat the air in the intake manifold or in the precombustion chamber. The glow plug is a heating element. Figure 1.17 illustrates one located in an engine. Unfortunately, the battery supplies the energy for heating. In some cases two batteries are provided, one for starting and one for heating. This takes some time, of course, and usually there is a pilot light on the dash to indicate that adequate preheating has occurred. Glow plugs used to take a minute or more to heat the precombustion chamber, but new designs provide heating in 6 to 8 seconds.

Another method sometimes used is to spray a starting fluid into the intake manifold. This method can cause engine damage if too much starting fluid is supplied, causing extreme pressure increases when it ignites, virtually an explosion. Starting fluid should never be used on engines with glow plugs, as engine damage can occur due to explosion-type detonation.

‖‖‖‖‖‖‖‖‖‖‖‖‖‖‖‖‖‖‖ ‖‖

Troubleshooting

Troubleshooting a diesel engine can be viewed as solving problems involving three general categories: trouble with engine operation (e.g., the engine not starting), trouble due to abnormal engine operating conditions while running, and trouble encountered while the engine is being taken apart for routine inspection. Troubleshooting begins by gathering all information observed and information supplied by the operator of the engine. Based on this information the troubleshooter proceeds in a logical fashion through the various symptoms of engine malfunctions. Usually, the engine problem does not result in one specific symptom, but in two or more symptoms, all of which must be analyzed to lead to the correct problem and its solution. For example, a smoky exhaust may be caused by engine overload, high lube oil consumption, injector malfunction, or some combination of these. By looking at other operating parameters, such as fuel or lube oil consumption, an exact diagnosis as to why the engine is smoking may be developed. Troubleshooting is a deduction process, a method of solving a mystery based on certain clues.

The engine manual provides a great deal of assistance in troubleshooting a specific engine, as malfunctions indicative of one manufacturer's engines are given attention, as well as those indicative of all engines. Once all information has been gathered, the best approach is to check the troubleshooting charts and proceed by doing the easiest things first. If there are five clues, it may be possible that the one which may be the most easily checked is the one causing the problem, or it may be associated with the most difficult to check. Start with the easiest and hope for the best.

Along with doing the easiest thing first, do not leap to conclusions when starting the troubleshooting task. Gather all information, take your time, think about what has happened, and then proceed in a logical fashion, one step at a time, eliminating various symptoms and finally finding the cause of the problem.

Following is a troubleshooting chart which includes symptoms, causes, and remedies for a variety of engine malfunctions. This troubleshooting guide applies to all engines, not to any one specific engine. The problems listed are those that occur most commonly in engine operation.

Troubleshooting Chart

Symptom	Cause	Remedy
Engine fails to start		
Electric starter	Discharged battery	Charge battery
	Defective starting motor or starter solenoid	Check continuity of starter system
	Poor electrical battery or starter connection	Remove and clean connections
Hydraulic	Low hydraulic oil pressure in accumulator	Pump up hydraulic pressure
Other systems	Fuel shut off	Open proper valves from fuel tank
	Fuel tank empty	Refill tank
	Fuel injection system air bound	Prime fuel system
	Clogged air filter	Replace air filter
	Engine cranks over too slowly	Check for weak battery
	Improper fuel metering pump timing	Check pump timing
	Loose high-pressure-fuel-line connections from injection pump to injector	Tighten connection
	Improper intake and exhaust valve clearances	Readjust valve clearance
	Low compression	Inlet and exhaust valves may not be seating properly; look for stuck piston rings, check position of compression release; cylinder head gasket may leak
Hard starting	Fuel pump needs priming	Prime system
	Air in system	Prime system
	Incorrect fuel pump timing	Check pump timing
	Throttle not set to starting fuel	Move throttle to start position
	Defective injectors	Test and adjust or replace
	Water in fuel system	Check system for water
	Leaky intake or exhaust valves	Check engine compression
	Improper valve clearance	Check intake and exhaust valve clearances
	Engine too cold	
Engine does not come up to speed or develop full power	Clogged air filter	Replace air filter
	Clogged fuel filters	Check and replace

(Continued)

(Troubleshooting Chart)

Symptom	Cause	Remedy
	Insufficient oil pressure at fuel injection pumps	Check to see if booster pump is operating; check pressure regulator; check for leaking valves
	Improper fuel injector pump timing	Check fuel pump timing
	Water in fuel	
	Air in fuel injection system	Bleed system to injectors
	Low compression	Check cylinder compression pressure
	Governor not functioning properly	Check governor adjustments
	Injectors are defective	Test injectors and adjust or replace
	Cylinders misfiring	Check for air in system, water in system, valve clearances; faulty injector; loose high-pressure-fuel-line connection
	Collapsed or clogged exhaust muffler	Check exhaust back pressure
	Poor-quality fuel	
	Too much friction	Bearings too tight; check lubrication; check for overheating, metal discoloration
	Engine overloaded	
Engine running unevenly	Improperly balanced fuel injection system	Check metering pump fuel rack settings
	Defective injector	Test injectors and adjust or replace
	Malfunctioning hydraulic governor	Adjust compensating needle
	Air leaks in booster or transfer suction line	Check for air leakage in fuel system
	Valves sticking	Use valve-freeing fluid until source of problem found
	Broken valves	
	Water in fuel	
	Clogged air filter	
	Two-cycle engines—check scavenging air valves, carbon deposits on exhaust ports	
Engine knocks	Early injection timing	Check and adjust timing
	Defective fuel injectors	Test injectors
	Poor-quality fuel	

(Troubleshooting Chart)

Symptom	Cause	Remedy
	Engine overload	Remove some of the engine load
	Inlet or exhaust valve sticking	
Engine overheats	Lack of coolant in system	Check radiator or expansion tank
	Fan belt loose	Check fan belt tension and adjust
	Radiator or coolers are dirty	Clean and replace
	Lack of lubricating oil causing friction	Check oil sump
	Worn water pump	Replace pump
	Engine overload	Remove excessive load
	Late fuel injection timing	Check injection fuel pump timing
	Faulty thermostat or thermostatic valve	Replace thermostat or adjust valve
Excessive lubricating oil consumption	External engine oil leaks	Locate and replace gaskets, etc.
	Excessive valve guide clearances	Replace valve guides
	Worn rings	Check compression pressure
	Excessively worn cylinder liners	Check compression pressure
	Excessive main and connecting rod bearing clearances	Check bearing clearances
	Excessive high oil temperature	Clogged or dirty oil cooler
Low lubricating oil pressure	Dirty or clogged oil filters	Check and replace
	Fuel oil dilution	Locate internal fuel leaks and change lube oil
	Faulty oil pressure regulator	Check and reset
	Worn lube oil pump	
	Too light an oil for high temperature operation	Change to heavier grade
	Excessively worn main and connecting rod bearings	
Black smoke	Engine overload	
	Malfunctioning injectors	Test, repair, or replace
	Improper fuel injection timing	Check injector timing
	Unbalanced fuel injection pumps	Check rack settings for each pump

(Continued)

(Troubleshooting Chart)

Symptom	Cause	Remedy
	Dirty oil filter	Replace filter
	Low engine compression	Run cylinder compression pressure test
Blue smoke	Excessive lube oil in combustion space	
	Worn valve guides	Replace guides or seats
	Worn oil control rings	Replace rings
	Worn cylinder liner	Replace liner
	Excessive main and connecting rod bearing clearances	Replace bearing
White smoke	Engine misfiring	
	Air in fuel injection system	Prime fuel injection system
	Water in fuel	Drain and replace fuel
	Bad injector	Test, repair, or replace

Glossary

ABDC: Abbreviation for after bottom dead center.

Abrasion: The wearing down of a surface by friction.

Acceleration: The rate of increase of velocity per unit time.

Accumulator: A device used for storing liquid under pressure; sometimes it smoothes out pressure surges in a hydraulic system.

Actuator: A device that uses fluid power to produce mechanical force and motion.

Additive: A substance that is added to improve another substances characteristics, such as in lubricating and fuel oils.

Advance: To cause to happen earlier, such as in the timing of injection pumps.

Aftercooler: A device used on turbocharged engines to cool air that has undergone compression, sometimes called an intercooler:

Air binding or locking: The presence of air in a pump or pipe which prevents the delivery or flow of a liquid.

Air cell: A space provided in the piston or cylinder to trap air during the compression stroke. It later flows out into the combustion chamber during the combustion process.

Air cleaner: A mechanism or device for filtering and removing dust, moisture, and other foreign matter from the air before it reaches the engine.

Air/fuel ratio: The ratio by weight or volume between air and fuel.

Air pollution: Contamination of the earth's atmosphere by pollutants such as smoke and toxic gases.

Align: To bring two or more components of a unit into the correct position relative to one another.

Allowance: The difference between the minimum and the maximum dimensions for proper functioning of the part.

Alloy: A mixture of two or more different metals, producing improved properties.

Alternator: An electromechanical device that produces ac power.

Ambient temperature: The temperature of the surrounding air.

Anneal: To strengthen metals by heating and then slowly cooling them.

Annular: In the form of an annulus, or ring-shaped.

Annulus: A figure bounded by concentric circles or cylinders.

Antifreeze: A chemical added to the coolant to lower its freezing point.

API gravity: A scale adopted by the American Petroleum Institute to identify the specific gravity of various fuels and oils.

Aspirated: To draw into or form a cylinder by suction produced by the piston in that cylinder.

ATDC: Abbreviation for after top dead center.

Atmosphere: The mass of gases surrounding the earth.

Atmospheric pressure: The pressure exerted by the atmosphere on the earth's surface, averaging 14.7 psi at sea level.

Atomizer: A device that disperses liquid, such as fuel, into fine particles.

Babbit: An antifraction metal used to line bearings, thereby reducing the friction between the moving parts.

Backfire: The ignition of a fuel–air mixture in the exhaust manifold.

Backlash: The distance between two surfaces before they contact, most often in reference to gears.

Back pressure: A resistance to the flow caused by a pressure exerted in the opposite direction to the flow.

Baffle: A device that slows down or diverts the flow of fluids.

Ball bearing: A bearing using steel balls as its rolling element between the inner and outer ring called a race.

Barometer: An instrument that measures atmospheric pressure.

BBDC: Abbreviation for before bottom dead center.

BDC: Abbreviation for bottom dead center.

Bearing: The contacting surface on which a revolving part rests.

Bearing clearance: The distance between the shaft and the bearing surface.

Bendix-type starter drive: A type of starter drive that causes the gear to engage when the motor starts rotating and to disengage automatically when it stops.

Bernoulli's principle: Given a fluid flowing through a tube, any constriction or narrowing of the tube will create an increase in the fluid's velocity and a decrease in its pressure.

Blow-by: Exhaust gases that escape past the piston rings and enter the crankcase.

Blower: A low-pressure air compressor, usually of a rotary or centrifugal.

Bore: The inside diameter of a cylinder.

Brake horsepower (bhp):   The usuable power delivered by the engine.

Brazing: The fastening of two pieces of metal together by heating the edges and then melting brass or bronze on the area, joining the pieces.

Breather pipe: A pipe opening into the crankcase to assist ventilation.

British gallon (Imperial gallon):   A volume measurement of 277.4 cubic inches.

British thermal unit (Btu):   Approximately the heat required to raise one pound of water one degree Farenheit in temperature.

Brush: The pieces of carbon of copper that make a sliding contact against the commutator or slip rings of a motor or generator.

BTDC: Abbreviation for before top dead center.

Buoyancy: The upward or lifting force exerted on a body by a fluid.

Bushing: A metallic or synthetic lining for a hole which reduces or prevents abrasion between two components.

Butterfly valve: A valve formed by a rotating disk in a pipe or cylindrical opening.

Bypass filter: An oil filter that filters only a portion of the oil flowing through the engine lubrication system.

Bypass valve: A valve that opens when the set pressure is exceeded. This allows the fluid to pass through an alternative channel.

Calibrate: To make an adjustment to an instrument so that it will accurately measure.

Calipers: A tool for measuring diameters, usually having curved legs and resembling a pair of compasses.

Calorie: Approximately the amount of heat required to raise one gram of water from 17°C to 18°C.

Cam: A rotating component of irregular shape. It is used to change the direction of motion from rotating to reciprocating.

Cam follower (valve lifter):   A part held in contact with the cam and to which the cam motion is imparted and transmitted to the push rod.

Cam noise: That portion of the cam that holds the valve wide open. It is the high point of the cam.

Camshaft: A shaft with cam lobes.

Camshaft gear: The gear that is fastened to the end of the camshaft and is driven directly or through intermediary gears by the crankshaft gear.

Carbon dioxide (CO_2):   A colorless, odorless gas which results when carbon is completely burned.

Carbon monoxide (CO):   A colorless, odorless toxic gas which results from the incomplete combustion of carbon.

Case-hardened: To harden the outer surface of metal to a given depth, while leaving the inner portion soft to absorb shocks and allow bending.

Cavitation: The formation of vapor bubbles in a fluid.

Cells (battery):   The individual compartments in the battery which contain positive and negative plates suspended in an electrolyte.

Celsius: A thermometer scale where water freezes at 0°C and boils at 100°C at atmospheric pressure.

Centrifugal force: A force exerted on a rotating object in a direction outward from the center of rotation.

Centrifugal pump: A pump using centrifugal force produced by a rapidly rotating impeller to move and increase a fluid's pressure.

Centrifuge: A device with a rapidly rotating bowl which separates the impurities of a fluid by high centrifugal force.

Cetane number: The rating of a diesel fuel's ignition quality.

Check valve: A valve that permits fluid to flow in only one direction.

Circumference: The distance or perimeter of a circle, 3.1416 times the diameter.

Clearance: The space between two components.

Clearance volume: The volume remaining above the piston when it is at TDC.

Closed cooling system: A cooling system that is not exposed to the atmosphere.

Clutch: A device used to connect and disconnect the power input and the power output.

Combustion: The burning or oxidation of a substance.

Combustion chamber: The chamber in which combustion in a diesel engine mainly occurs.

Compound: A chemical combination of two or more elements.

Compressed air: Air that is at a pressure greater than atmospheric pressure.

Compressibility: The characteristic of a substance whereby its density increases as the pressure acting on it increases.

Compression: Increasing the pressure of a gas, typically by reducing its volume.

Compression check: A measurement of the compression pressure of each cylinder as the engine is running or being turned over at firing speed.

Compression ignition: The ignition of fuel caused by the high temperature of the air it is injected into.

Compression pressure: The pressure in the combustion chamber at the end of the compression stroke without fuel being added.

Compression ratio: The ratio between the volume left in the cylinder at top dead center divided by the volume in the cylinder when the piston is at bottom dead center.

Compression release: A device to prevent the intake or exhaust valves from closing completely, thereby permitting the engine to be turned over without compression.

Compression ring: The piston rings used to reduce combustion gas leakage into the crankcase, located on the uppermost part of the piston side.

Compressor: A mechanical device used to increase the pressure of a gas, such as air.

Concentric: Having the same center.

Condensation: The process of a vapor becoming a liquid.

Conduction: The transmission of heat through a subtance without the substance moving and caused by a temperature difference across the substance.

Connecting rod: The rod joining the piston and the crankshaft.

Connecting rod bearing: The bearing used to join the connecting rod and the crankshaft.

Control: To regulate or govern the output of a device.

Convection: The transfer of heat from or to a solid to a liquid caused by a temperature difference between the solid and liquid.

Coolant: A liquid used in a cooling system.

Corrode: The dissolution of a substance by chemical action.

Counterbore: A cylindrical enlargment of the end of a cylinder bore.

Countersink: To cut or shape a depression in an object so that the head of a screw or bolt may set flush or below the surface.

Counterweights: The weights that are mounted on the crankshaft opposite each crank throw. These reduce the vibration caused by the crank webs by creating a counteracting force.

Coupling: A device used to connect two rotating components, typically shafts.

Crankcase: The casing that surrounds the crankshaft.

Crankcase scavenging: A scavenging method using the pumping action of the power piston in the crankcase to provide scavenging air.

Crankpin: That portion of the crankshaft that is attached to the connecting rod.

Crankshaft: A rotating shaft for converting reciprocating motion into rotary motion.

Crankshaft gear: The gear attached to the end of crankshaft that drives the camshaft.

Crank throw: The distance of one crankpin with its two webs.

Critical compression ratio: The lowest compression ratio at which any particular fuel will ignite by compression ignition.

Critical speeds: The speeds at which the frequency of the power strokes synchronize with the crankshaft's natural frequency. This causes severe vibration of the crankshaft and may cause its failure.

Crude oil: Petroleum as it comes from the ground.

Crush: A deliberate distortion of an engine's bearing shell to hold it in place during engine operation.

Cylinder: The piston chamber of an engine.

Cylinder block: The solid casing which includes the cylinder and water jackets (cooling fins for air-cooled engines).

Cylinder head: The replaceable portion of the engine that seals the cylinder at the top.

Cylinder liner: A sleeve that is inserted in the bores of the engine block which make up the cylinder wall.

Dead center: Either of the two positions when the crank and connecting rod are in a straight line at the end of a stroke.

Deceleration: The opposite of acceleration—a decrease in velocity with time.

Deflection: Bending or movement away from the normal position due to loading.

Density: The weight per unit volume of a substance.

Detergent: A chemical with cleaning qualities that is added to the engine oil.

Detonation: The rapid burning of a portion of the fuel in the combustion chamber, causing high-pressure waves.

Diaphragm: Any flexible division separating two compartments.

Dipstick: A device to measure the quantity of oil in the reservoir.

Direct-cooled piston: A piston that is cooled by the internal circulation of a liquid.

Displacement: The volume swept by all of the pistons in the engine in making one stroke.

Distillation: Heating a liquid and then condensing the vapors produced by the heating process.

Double-acting: A device that is able to perform its task while moving in both directions.

Dowel: A pin, usually of cylindrical shape, used to fasten an object in position.

Dribbling: Unatomized fuel running from the fuel injector nozzle.

Drop-forged: Formed by hammering and forcing into a desired shape.

Dry liner: A cylinder liner which is supported its entire length by the cylinder wall. The coolant does not come in contact with the liner.

Dynamic balance: Balancing a device while it is rotating so that there is no vibration.

Eccentric: Circles that do not have the same center.

Efficiency: Output divided by input; for an engine, the power produced by the fuel energy supplied.

Emulsify: To suspend oil and water in a mixture where the two do not readily separate.

End play: The amount of axial movement in a shaft.

Engine displacement: *See* Displacement.

Excess air: The percent of air above the minimum required for complete combustion in an ideal process.

Exhaust gas: The products of combustion in an engine.

Exhaust manifold: A device that connects all the exhaust ports to one outlet, the exhaust pipe.

Exhaust port: The opening through which exhaust gas passes from the cylinder to the manifold.

Exhaust valve: The valve which, when opened, allows exhaust gas to leave the cylinder.

Eye bolt: A bolt threaded at one end and bent in a closed loop at the other end.

Fahrenheit: A temperature scale where the freezing point of water is 32°F and the boiling point of water is 212°F at atmospheric pressure.

Fatigue: Deterioration of a material's strength caused by constantly changing stress or forces acting on it.

Feeler gauge: A strip of steel ground to a precise thickness and used to check clearances.

Filter: A device used for cleaning or purifying a liquid or gas.

Fire point: The lowest temperature at which an oil heated in a standard apparatus will ignite and continue to burn.

Firing order: The order in which the cylinders deliver their power stroke.

Firing pressure: The highest pressure reached in a cylinder during the combustion process.

Fit: The closeness between the surfaces of two machined components.

Flange: A metal part that is spread out like a rim and used to join one device, often a pipe, to another device.

Flare: To open or spread outwardly. A method used in joining tubing fittings.

Flash point: The temperature at which a substance, usually a liquid, will give off a vapor that will flash or burn momentarily when ignited.

Fluid: A gas, a liquid or combination of the two.

Flywheel: A rotating device used for storing energy in order to carry the pistons through the compression process and to minimize cyclicial speed viariations.

Foot-pound: The amount of work accomplished when a force of one pound moves something one foot in distance.

Forged: Shaped with a hammer, usually heated first.

Foundation: The structure on which an engine is mounted.

Frinction: The resistance to motion due to contact between two surfaces.

Fuel mixture: A ratio of fuel and air.

Fulcrum: The pivot point of a lever.

Full-floating piston pin: A piston pin free to turn in the piston boss where it connects to the connecting rod.

Full-flow oil filter: A filter that has all the engine oil passing through it before entering the engine parts.

Gallery: Passageways inside a wall or casting.

Galvanic action: When two dissimilar metals are immersed in certain solutions a flow of electrons will occur and cause metal deterioration.

Gasket: A layer of material used to seal machined surfaces, preventing leakage.

Gauge pressure: Pressure above atmospheric pressure.

Gear-type pump: A pump that uses the spaces between the adjacent teeth of gears for moving the liquid.

Gland: A device used to prevent the leakage of gas or liquid along a shaft.

Glaze: A smooth, glassy surface.

Glow plug: A heater plug for the combustion chamber using electrical heating.

Governor: A device for controlling the speed of an engine.

Gravity: The force that tends to draw all bodies toward the center of the earth.

Heat: A form of energy flow caused by a temperature difference between two bodies.

Heat exchanger: A device used to heat or cool by transferring heat from one substance to another.

Heating value: The amount of heat produced by burning one pound of fuel.

Hone: A tool with an abrasive stone used for removing metal; to hone is the act of removing metal with a hone.

Horsepower (hp):   A unit of power equal to 33,000 foot-pounds of work performed in one minute.

Hunting: Rhythmic overspeeding and then underspeeding, caused by a governor instability.

Hydraulic governor: A governor using a liquid, usually oil, to operate the fuel control linkage.

Hydrocarbon (HC):   A compound formed by hydrogen and carbon.

Idling: An engine running without load at the lowest speed possible without stalling.

Ignition: The start of the combustion process.

Ignition lag: The time between the start of injection and ignition.

Immersed: To be completely covered by a liquid.

Indirectly cooled piston: A piston cooled primarily by heat conduction through the cylinder walls. The heat is transferred from the piston to the piston rings to the cylinder walls.

Inertia: The property of matter that causes it to tend to remain at rest if not moving or to move in a straight line if already moving.

Inhibitor: Any substance that retards or prevents chemical reactions such as corrosion or oxidation.

Injection pump: A high-pressure pump that delivers a variable amount of fuel to the combustion chamber.

Injector: A device used to force fuel into the combustion chamber.

In-line engine: An engine in which all cylinders are in a straight line.

Intake manifold: A casting that joins the intake ports or valves with the air filter or turbocharger.

Intake valve: The valve that allows air to enter the cylinder.

Intercooler: A heat exchanger located between the turbocharger air discharge to remove the heat of compression before the air enters the cylinder.

Isochronous governor: A governor having zero speed droop.

Journal: That portion of a shaft that rotates in a bearing.

Key: A fastening device used to join two concentric shafts. A slot is formed in each shaft, the key fits in the slot, and the shafts rotate together.

Keyway: The groove cut in the shaft to hold the key.

Kilometer: A unit of length equal to 0.6215 mile.

Kilowatt: A unit of power equal to horsepower.

Kinetic energy: The energy an object has due to its velocity.

Knocking: A sharp pounding noise in the cylinder caused by too rapid a pressure increase in the combustion process.

Land: The projecting portion of a grooved surfaced; for example, piston rings rest on ring lands cut in the piston.

Lap: A method of polishing a metal surface by rubbing it with a very fine grinding compound.

Liner: An insert put into a cylinder to wear instead of the cylinder wall; also called a sleeve.

Linkage: A movable connection between two devices.

Liter: A volume measurement equal to 0.2642 gallon.

Lubricant: A substance that decreases the friction between two surfaces.

Lug: A condition of engine operation that occurs when the engine operates at a speed below its maximum torque speed.

Main bearing: A bearing that supports the crankshaft.

Mechanical advantage: Using a lever to increase the effect of a force in moving an object. Specifically, the distance through which the force is exerted divided by the distance the weight is raised.

Metering fuel pump: A fuel pump that delivers a controlled amount of fuel per cycle.

Misfiring: When one cylinder does not have the combustion process occur at the same time and the same pressure rise as the other cylinders.

Needle bearing: A roller-type bearing in which the roller bearing diameter is smaller than its length.

Neoprene: A synthetic rubber highly resistant to oil, heat, and oxidation.

Nitrogen oxides (NO_x):   The compounds formed at high temperature by nitrogen and oxygen.

Nozzle: The end of the injector, which contains orifices through which fuel is injected into the cylinder.

Ohm: A unit of electrical resistance.

Oil bath air cleaner: An air filter that uses the principle of air flowing through oil to remove impurities in the air.

Oil cooler: A heat exchanger used for lowering the temperature of engine lubricating oil.

Oil filter: A device used to remove solid impurities from oil.

Oil gallery: A drilled or cast passage in the cylinder head block and crankcase which carries oil for lubrication and cooling.

Oil seal: A mechanical device on a shaft which prevents oil leakage around the shaft.

Orifice: An opening in a solid disk through which fluid flows.

Overhead camshaft: A camshaft that is located above the cylinder head.

Overrunning clutch: A clutch mechanism that transmits power in one direction only.

Overrunning clutch starter drive: A mechanical device that locks in one direction but turns freely in the opposite direction.

Overspeed governor: A governor that shuts off the flow of fuel to the engine when excessive speed is reached.

Oxidation: The chemical reaction in which oxygen unites with another substance.

Packing: A flexible material, most often stringlike, that is used to fill the space between two parts which move relative to one another.

Paper air filter: An air filter with a special pleated paper element through which the air passes and is filtered.

Penetrating oil: An oil with properties that aid in the removal of parts that are rusted together.

Periphery: The outside boundary of an object.

Pilot shaft: A shaft position in a component as a means of aligning the component with another piece of equipment.

Pintle-type nozzle: A type of fuel injection nozzle that produces a conical spray of oil.

Piston: A movable cylinder located in a cylinder block and attached to the connecting rod.

Piston boss: The reinforced area around the piston pin bore.

Piston displacement: The volume swept by the piston in moving from bottom dead center to top dead center.

Piston head: The portion of the piston above the top ring.

Piston lands: The space on the side that is between the ring grooves.

Piston pin (wrist pin):   A cylindrical pin that passes through the piston bore and is used to connect the connecting rod to the piston.

Piston ring: A split ring placed in a groove of the piston and used to seal the space between the piston and the wall.

Piston skirt: The portion of the piston that is below the piston bore.

Piston speed: The total distance traveled by each piston in one minute.

Pivot: The pin or shaft on which a component moves.

Play: The movement (distance) between two components.

Pneumatics: The study of flow and pressure in gases.

Polar timing diagram: A graphical representation of the events occurring in an engine cycle in relation to crankshaft rotation.

Ports: Openings in the cylinder block and head for the flow of oil, coolants, air, and exhaust.

Port scavenging: The introduction of scavenging air into the cylinder by the piston uncovering ports in the cylinder wall.

Pour point: The lowest temperature at which an oil will flow.

Power: The rate of doing work.

Precombustion chamber: A portion of the combustion chamber that is connected to the cylinder through a small opening. Fuel is injected here and flows into the main combustion chamber to complete the process.

Pressure: The force per unit area.

Pressure cap: A special radiator cap with pressure and vaccuum relief valves.

Pressure lubrication: A lubrication system where oil is brought under pressure to the points that need lubrication.

Pressure relief valve: A valve that opens to limit the maximum pressure in a system. It closes once the pressure drops below the opening valve.

Printed circuit: An electrical curcuit that is made by printing or pressing the conductor on an insulating material.

Pump: A device that is used to increase the pressure of and move liquids.

Pumping loss: The power used to replace the exhaust gas in a cylinder with fresh air.

Pushrod: A cylindrical rod or tube used to transmit motion from the cam shaft to the rocker arm.

Pyrometer: A temperature indicator.

Race (bearing):   The inner and outer groove or channel that holds the bearings.

Raceway: The surface of the race.

Radial clearance: The distance between a shaft and bearing measured in the radial direction (peripendicular to the shaft).

Radiator: A heat exchanger in the cooling system that has a flow of heat between the surrounding air and the system coolant.

Rebore: To bore out a cylinder to a diameter larger than the original.

Reciprocating motion: A back-and-forth or up-and-down movement.

Relief valve: *See* Pressure relief valve.

Retard: To cause the timing of an event to occur later in the cycle.

Reverse flush: To pump water or a cleaning agent through a cooling system in the direction opposite to normal flow.

Ring expander: A metal strip or spring that is placed behind the piston oil control rings to force them against the cylinder wall.

Ring groove: A groove machined in the piston side wall to hold the piston ring.

Rocker arm: A lever device that transmits the motion of the camshaft to the valve stem.

Rocker-arm shaft: The shaft on which the rocker arms pivot.

Roller bearing: An antifriction bearing using cylindrical or tapered bearings spaced in an inner and outer ring.

Roots blower: An air blower similar to the gear-type pump.

Running fit: A machine fit with sufficient clearance for expansion and lubrication.

SAE: Abbreviation for the Society of Automotive Engineers.

SAE viscosity numbers: A simplified viscosity rating of oils based on Saybolt viscosity.

Safety factor: The percent of extra strength or capacity that a device has above the designed minimum requirement.

Saybolt viscosimeter: A calibrated container for determining an oils viscosity.

Scale: Precipitated material caused by minerals and salts in water.

Scavenging: The displacement of exhaust gas from the cylinder by fresh air.

Scraper ring: An oil control ring that removes oil from the distributes oil on the cylinder wall.

Sealed bearing: A bearing that is lubricated and sealed at the factory and cannot be lubricated while in service.

Sediment: Solid material in a liquid.

Semifloating piston pin: A piston pin that is clamped either to the connecting rod or the piston boss.

Shaft horsepower: The power delivered by the engine's crankshaft.

Shim: Thin, flat pieces of metal, usually steel or brass, used to increase the distance between two components.

Shrink-fit: A tight fit between two components caused by heating one so that it expands and slides over the other; on cooling it contracts and fits tightly around the other.

Shroud: The enclosure around a fan or engine that directs the flow of air to it.

Silencer: A device used to reduce the noise in the exhaust or intake system.

Sludge: Deposits caused by the action of oil, water, and dirt.

Snap ring: A fastening device in the shape of a split ring that it rides in a groove in a bore, such as holding the piston pin in the piston boss.

Sodium valve: A valve designed with a hollow stem that is filled with metallic sodium; this improves the heat transfer from the valve.

Solenoid: An electromagnetic valve or switch.

Specific gravity: The ratio of the weight per given volume of a substance divided by the weight of water in the same volume.

Spline: The land between two grooves on a shaft.

Stability: The resistance of a fluid, not the change in its chemical composition; also, the ability of a governor to control an engine at a given speed or load.

Stroke: The distance the piston moves in going from bottom dead center to top dead center.

Stroke-to-bore ratio: The length of the stroke divided by the diameter of the cylinder.

Suction valve: Intake valve.

Sump: A receptacle into which liquid drains.

Supercharger: An air compressor driven by the engine which supplies air to cylinders under higher than atmospheric pressure.

Surge: A momentary rise, then fall, of pressure or speed.

Synchronize: To cause two or more events or occurrences to happen at the same time.

Tachometer: An instrument used to indicate rotating speed, most often engine crankshaft speed.

Tap: A tool used for cutting threads in a hole.

Tappet: The rocker arm.

TDC: Abbreviation for top dead center.

Temper: Increasing the hardness of a metal by heating it and suddenly cooling it.

Tension: A stress on a material caused by stretching it.

Thermocouple: The part of a pyrometer that consists of two dissimilar metal wires fused together at one end. The free electrons at this junction creates a voltage potential that is proportional to the temperature of the junction; the voltage is measured and converted to temperature.

Thermometer: An instrument for measuring temperature.

Throttling: Reducing the flow and pressure of a fluid.

Throw: The part of a crankshaft to which the connecting rod is fastened.

Thrust bearing: A bearing or washer that prevents axial movement of a shaft.

Timing gears: The gears attached to the crankshaft and other components, such as the camshaft, that coordinate their rotation with the crankshaft.

Timing marks: The marks located on the flywheel or vibration damper to check injector timing.

Tolerance: The variation from size specifications that a part may have.

Torque: A force that produces a twisting movement.

Torsional vibration: The vibration due to the twisting and relaxing of a shaft.

Turbine: A series of curved vanes attached to a shaft or rotor which move due to the pressure of a liquid or gas flowing across them.

Turbocharger: A turbine driven by exhaust gases and directly attached to a rotating compressor that compresses the intake air to the cylinder.

Turbulence chamber: A combustion chamber connected to the cylinder by a small passageway.

Two stroke cycle: An engine cycle in which all events, intake, compression power, and exhaust occur in one revolution of the crankshaft or two strokes of the piston.

Uniflow scavenging: A scavenging method where air enters from ports at one end of the cylinder and exhaust leaves at the other end.

Unit injector: A combined fuel injection pump and fuel nozzle.

Vacuum: A pressure that is less than atmospheric pressure.

Valve: A device that is used to allow or prevent the flow of a liquid or gas.

Valve duration: The time, measured in degrees of crankshaft rotation, that a valve remains open.

Valve float: A condition where the valves are forced open due to valve spring vibration or engine vibration at a given speed.

Valve guide: A hollow-sized shaft that is inserted in the cylinder head to maintain valve stem alignment and wear instead of the cylinder head.

Valve keeper: A device that locks the valve spring retainer to the valve stem; also known as a valve retainer ring.

Valve lift: The distance a valve moves when going from the fully closed to fully open position.

Valve margin: The distance between the edge of the valve and the edge of the face.

Valve oil seal: A sealing device that prevents excess oil from entering the area between the stem and the valve guide.

Valve overlap: The period, measured in degrees of crankshaft rotation, during which both the intake and exhaust valves remain open.

Valve rotator: A mechanical device that is attached to the valve stem and rotates the valve stem for each movement of the rocker arm.

Valve seat: The surface on which the valve face rests when it is closed.

Valve seat insert: A hardened steel ring that is placed in the cylinder head which the valve seats against, wearing instead of the head.

Venturi: A specifically shaped tube or restriction that is used to increase velocity and decrease pressure in a fluid. It is often used in measuring the flow of the fluid.

Vibration damper: A device that is mounted to the front of the crankshaft to reduce the torsional vibration of the crankshaft.

Viscosity: The property of a liquid or gas that indicates its resistance to flow.

Volmetric efficiency: An efficiency relating how much air is actually drawn into a cylinder to the maximum amount that may be drawn into it.

Water jacket: The enclosed passages in the cylinder block and head that direct the flow of the cooling water through the engine.

Wet liner: A cylinder liner or sleeve that is exposed directly to the cooling water.

Yoke: A Y-shaped link that directs mechanical action to two points.

Index

A

Accumulator, 93
Additives, 79
Aftercooler, 86
Air cleaner:
  dry type, 83
  oil–bath type, 83
Air–fuel ratio, 17
Air–intake system, 83
Air starting system, 94
Antifreeze, 71, 75

B

Back pressure, 89
Batteries, 91
Bearings, 36
Block, 1
Blow–by, 79
Blowers, 88
Bore and stroke, 9
Bottom dead center, 6
Brake horsepower, 24
British thermal unit (Btu), 24
Bushing, 35

C

Cam followers, 41
Camshaft, 4, 40
Camshaft gears, 40
Camshaft lobes, 4, 40
Casting, 1
Cetane number, 46
Clearance volume, 7
Combustion, 13, 16

Combustion chambers, 17
  energy cell, 20
  open, 17
  precombustion, 19
  turbulence, 19
Compression pressure, 13
Compression ratio, 7
Compression stroke, 8
Connecting rod, 35
  bearings, 36
Cooling system, 70
Corrosion, 75
Counterbore, 31
Counterweights, 39
Crank angel, 11
Crankshaft, 2, 37
  main bearing, 38
Cycles, 6
Cylinder block, 1
Cylinder components, 9
Cylinder head, 3
Cylinder liner (sleeve), 2, 29

D

Delay period, 12
Diesel fuel, 3
  properties, 47
Displacement, 6
Dry-type air cleaner, 83

E

Electric starting system, 91
Electronic fuel injection, 67
Engine cycles, 6
Engine parts, 1
Erosion, 75